Enabling Technologies for the Successful Deployment of Industry 4.0

Manufacturing Design and Technology Series

Series Editor
J. Paulo Davim, University of Aveiro, Portugal

This series will publish high-quality references and advanced textbooks in the broad area of manufacturing design and technology, with a special focus on sustainability in manufacturing. Books in the series should find a balance between academic research and industrial application. This series targets academics and practicing engineers working on topics in materials science, mechanical engineering, industrial engineering, systems engineering, and environmental engineering as related to manufacturing systems, as well as professions in manufacturing design.

Drills
Science and Technology of Advanced Operations
Viktor P. Astakhov

Technological Challenges and Management
Matching Human and Business Needs
Edited by Carolina Machado and J. Paulo Davim

Advanced Machining Processes
Innovative Modeling Techniques
Edited by Angelos P. Markopoulos and J. Paulo Davim

Management and Technological Challenges in the Digital Age
Edited by Pedro Novo Melo and Carolina Machado

Machining of Light Alloys
Aluminum, Titanium, and Magnesium
Edited by Diego Carou and J. Paulo Davim

Additive Manufacturing
Applications and Innovations
Edited by Rupinder Singh and J. Paulo Davim

Emotional Intelligence and Neuro-Linguistic Programming
New Insights for Managers and Engineers
Edited by Carolina Machado and J. Paulo Davim

Business Intelligence and Analytics in Small and Medium Enterprises
Edited by Pedro Novo Melo and Carolina Machado

Enabling Technologies for the Successful Deployment of Industry 4.0
Edited by Antonio Sartal, Diego Carou, and J. Paulo Davim

For more information about this series, please visit: www.crcpress.com/Manufacturing-Design-and-Technology/book-series/CRCMANDESTEC

Enabling Technologies for the Successful Deployment of Industry 4.0

Edited by
Antonio Sartal, Diego Carou, and J. Paulo Davim

CRC Press
Taylor & Francis Group
Boca Raton London New York

CRC Press is an imprint of the
Taylor & Francis Group, an **informa** business

Published 2020
by CRC Press
6000 Broken Sound Parkway NW, Suite 300, Boca Raton, FL 33487-2742

and by CRC Press
2 Park Square, Milton Park, Abingdon, Oxon, OX14 4RN

First issued in paperback 2021

Publisher's Note
The publisher has gone to great lengths to ensure the quality of this reprint but points out that some imperfections in the original copies may be apparent.

ISBN 13: 978-0-367-15196-6 hbK)
ISBN 13: 978-1-03-224060-2 (pbk)

DOI: 10.1201/9780429055621

Typeset in Times
by Deanta Global Publishing Services, Chennai, India

Contents

Preface

Industry 4.0 is a general term for identifying all the changes that are being produced in industry as a consequence of the so-called Fourth Industrial Revolution. This movement is completely changing traditional industries, greatly affecting economic activity. It originates from a high-tech strategy of the German government around 2011. And, since then, its worldwide development has been exponential and many interchangeable designations for Industry 4.0, related to national and international strategies such as "smart factories" and "factories of the future", have appeared since the apparition of German term *Industrie 4.0*.

We would like to highlight the sentence coined by the famous consultant Warren Bennis a few years ago: "The factory of the future will have only two employees, a man and a dog. The man is there to feed the dog. The dog is there to keep the man from touching the equipment". Undoubtedly, this provocative definition of Industry 4.0 sums up quite well the enormous current uncertainty about this growing phenomenon. On the one hand, it seems that the possibilities associated with new technologies are being overvalued, at least as regards the short and medium term. As a consequence of this, there is also a huge fear in society as regards the role that humans will play in this new digital and technological environment: our only function will be to "feed the famous dog" as Mr. Bennis points out. On the other hand, although everyone, including managers and politicians, uses this term airily, many of them do not know in depth what it is really about, or what technologies should be under the Industry 4.0 umbrella. In fact, the real problem has been the banalization of the term 4.0 due to its indiscriminate use. Nowadays, this famous suffix 4.0 is added to any technology or process (new or not so new) and sold as something new, revolutionary, when in fact it could be essentially the same as always.

According to these two aforementioned issues, the objective of this book is to provide an accurate and clear vision about what this new paradigm consists of, explaining its purpose, the challenges to overcome and, above all, giving a detailed description of the various technologies that integrate it and its applications. From an exhaustive review of the literature and considering the latest advances in the industrial field, we present a synthesis of the main disruptive technologies that, according to the principles surrounding Industry 4.0 (e.g., interoperability, decentralized decisions, information transparency, etc.) would fit within this framework. Through a pleasant and understandable language for specialists and non-specialists, we also intend to identify and transmit the significance of the implications derived from this transformation at all levels, especially with regard to the economic, industrial, and social spheres.

Subsequently, a detailed presentation of the technologies that will enable this new scenario is described in the book. We intend not only to make a description of its main functionalities, but also to evaluate its capacity to generate true manufacturing strengths and finally sustainable competitive advantages over time, both theoretically and through different successful implementation cases. In particular, several technologies are considered as key enabling technologies for Industry 4.0. Among these technologies, artificial intelligence, big data and cloud computing, autonomous robots and cobots, etc. can be cited. Although most people in industry and even in society are familiar with the names of these new technologies, the way they work and their possible applications for them are not well-known. Particularly, it is important to note that these technologies can be considered as novel, some of them can be considered as disruptive and, thus, a lot of new developments and customer-driven applications in different sectors have appeared in a short time.

Although these technologies are commonly related to industry (mainly manufacturing), they can be transferred to other activity sectors. In fact, several applications of these novel technologies are shown in different sectors and environments (e.g., public vs. private) along the value chain, in small and medium-sized enterprises (SMEs), services vs. manufacturing and high-tech vs. low-tech and with insights for certain specific sectors such as health, entertainment, and education, among others.

Finally, the relationship between the organizational and technological sides of the organizations is analyzed in depth in the last chapter: beyond a purely technocentric vision, we try to investigate how the technologies serve as support for the organizational routines and the workers (and vice versa). Accordingly, this last chapter conducts an exhaustive review about the role that new I4.0 technologies play in the value chain. The various technologies are classified according to the impact that they have on responsiveness, that is, on the supply chain flexibility and/or response speed, assessing furthermore the potential implications for academia and management.

We invite the reader to accompany us on this "technological journey" and we also take the opportunity to thank all contributors to the book for their efforts and valuable contributions, as well as the support given by CRC Press throughout this process.

Editors

Antonio Sartal is currently a "Beatriz Galindo" Postdoc Researcher at the University of Vigo, Spain. He managed the Department of R&D of a food multinational for the past ten years, until he recently joined a research team working on technology management and organizational innovation. His research interests include the intersection of lean thinking, innovation management, and information technologies. He has published his work in the *Journal of Operations Management, Supply Chain Management: An Int. Journal, Journal of Cleaner Production, Int. Journal of Manufacturing Systems, Computer and Operations Research* and *IEEE Transactions On Engineering Management*, among others. He has been awarded the *IEOM (Industrial Engineering and Operations Management Society International) Young Researcher Award 2019* for his contributions in the field of industrial engineering and operations management.

Position and Affiliation: Postdoc Researcher, University of Vigo, Spain Facultade de Economía e Admón. de Empresas

Diego Carou received his PhD degree in Industrial Engineering from the National University of Distance Education (UNED), Spain in 2013 and Industrial Engineering degree from the Polytechnic University of Valencia, Spain in 2004. He has international postdoctoral experience in manufacturing processes at several European universities. He worked as Assistant Professor at the Department of Mechanical and Mining Engineering at the University of Jaén, Spain from 2017 to 2020. He currently works as Assistant Professor at the Department of Design in Engineering at the University of Vigo, Spain. He has published more than 40 contributions in journals, conferences, and book chapters. He also serves as Reviewer for several international journals and carries out editor activities. His current research interests include additive manufacturing, Industry 4.0, machining, and sustainable manufacturing.

Position and Affiliation: Assistant Professor, School of Aeronautic and Space Engineering, University of Vigo, Spain

J. Paulo Davim received his PhD degree in Mechanical Engineering in 1997, MSc degree in Mechanical Engineering (materials and manufacturing processes) in 1991, Mechanical Engineering degree (five years) in 1986, from the University of Porto (FEUP), Portugal, the Aggregate title (Full Habilitation) from the University of Coimbra, Portugal in 2005 and DSc degree from the London Metropolitan University, UK in 2013. He is Senior Chartered Engineer by the Portuguese Institution of Engineers with an MBA and Specialist titles in Engineering and Industrial Management, as well as in Metrology. He is also Eur Ing by FEANI-Brussels and Fellow (FIET) by IET-London. Currently, he is Professor at the Department of Mechanical Engineering at the University of Aveiro, Portugal. He has more than 30 years of teaching and research experience in Manufacturing, Materials, Mechanical, and Industrial Engineering, with special emphasis in Machining & Tribology. He also has an interest in Management, Engineering Education, and Higher Education for Sustainability. He has guided large numbers of postdoc, PhD, and master's students and has coordinated and participated in several financed research projects. He has received several scientific awards. He has worked as Evaluator of projects for the European Research Council (ERC) and other international research agencies as well as examiner of PhD theses for many universities in different countries. He is the Editor in Chief of several international journals, Guest Editor of journals, books Editor, book Series Editor, and Scientific Advisory for many international journals and conferences. Currently, he is an Editorial Board member of 30 international journals and acts as Reviewer for more than 100 prestigious Web of Science journals. In addition, he has also published as editor (and co-editor) more than 125 books and as author (and co-author) more than 10 books, 80 book chapters, and 400 articles in journals and conferences (more than 250 articles in journals indexed in Web of Science core collection/h-index 54+/9500+ citations, SCOPUS/h-index 59+/11500+ citations, Google Scholar/h-index 76+/19000+).

Position and Affiliation: Professor, University of Aveiro, Portugal

Contributors

Gülçin Büyüközkan
Department of Industrial Engineering
Faculty of Engineering and Technology
Galatasaray University
Istanbul, Turkey

Diego Carou
Department of Design in Engineering
University of Vigo
Vigo, Spain

J. Paulo Davim
Department of Mechanical Engineering
University of Aveiro
Aveiro, Portugal

Guido Guizzi
Department of Chemical and Materials
   Engineering and Operations
   Management
University of Naples Federico II
Naples, Italy

Öykü Ilıcak
Department of Industrial Engineering
Faculty of Engineering and Technology
Galatasaray University
Istanbul, Turkey

Vidosav D. Majstorovic
University of Belgrade
Faculty of Mechanical Engineering
Belgrade, Serbia

Juan Manuel Maqueira-Marín
Departamento de Organización de
   Empresas, Marketing y Sociología
Universidad de Jaén
Jaén, Spain

Pedro José Martínez-Jurado
Centro Universitario de la Defensa de
   Zaragoza
Zaragoza, Spain

Miguel Ángel Moreno
Department of Mechanical
   Engineering
University Carlos III of Madrid
Madrid, Spain

José Moyano-Fuentes
Departamento de Organización de
   Empresas, Marketing y Sociología
Universidad de Jaén
Jaén, Spain

Miguel Núñez-Merino
Departamento de Organización de
   Empresas, Marketing y Sociología
Universidad de Jaén
Jaén, Spain

Ercan Oztemel
Industrial Engineering Department,
   Engineering Faculty
Marmara University
Istanbul, Turkey

Roberto Revetria
Department of Mechanical, Energy,
   Logistics Engineering and
   Engineering Management
University of Genoa
Genoa, Italy

Anastasiia Rozhok
Department of Mechanical, Energy,
   Logistics Engineering and
   Engineering Management
University of Genoa
Genoa, Italy

and

Department of International Education
   and Scientific Collaboration
Bauman Moscow State Technical
   University
Moscow, Russia

Slavenko M. Stojadinovic
University of Belgrade
Faculty of Mechanical Engineering
Belgrade, Serbia

Deniz Uztürk
Department of Business
   Administration
Faculty of Economics and
   Administrative Sciences
Galatasaray University
Istanbul, Turkey

1 Understanding Digital Transformation

Ercan Oztemel

CONTENTS

1.1 INTRODUCTION

Digital transformation is defined as the changes associated with digital technology applications and integration of that to all aspects of human life. This transformation is in fact to move from physically empowered life to the digital one. Everyone should accept the fact that everything in the world is changing. The only thing that doesn't change is the change itself. This has been realized throughout history. When it began, human life was based on hunting animals. As a result of use of the seed, life was empowered by agriculture and land owners became powerful social actors. The change did not stop, and machines were invented just before the last century.

This consciously created the so-called "first industrial society" and landowners had to leave their social impacts and positions to bosses. Industrial revolutions are followed one after the other. Each transformation also led to new countries developing superiority over weak followers.

During the end of last century scientific progress experienced the term "information" as the key to success in business. Information management systems and methodologies completely changed the way of life and had a remarkable impact on nearly all aspects of human life. People became unable to live without the support of computers. Computer engineering and programming were considered the most important drivers of the society. Enterprises had to change their operating platforms from traditional machines to software-driven automated ones. It was no surprise to see the political powers of digital investors. Being able to manage information and progress along with the respective technologies, such as artificial intelligence over time, made it possible to deal with and well manage the "knowledge". This naturally led to the generation of intelligent systems. In recent years, people started to utilize knowledge-based systems. This in turn encouraged the transformation towards a knowledge-driven society.

It is now very obvious that the change will continue and the wisdom society will emerge. Wisdom managers will be the key power in future societies. Progress in the same vein should be followed very carefully, in order to cope with possible challenges in the most effective way. Not only should the nations take necessary actions, but the enterprises should also develop proper action plans and follow a systematic process in order to be aligned with possible transformation. This process should be continuously updated in accordance with the basic requirements of the change. This chapter intends to provide guidelines for those who have an interest in digital transformation and wish to comply with it.

1.2 TRANSFORMATION AND THE BASIC CHALLENGES

As it is clearly stated above, everything is changing. There is no way to stop this. The change appears from various aspects including the technological, methodological, environmental, managerial and customer related changes as well as those in manufacturing systems, etc. Oztemel (2010) analyzed each of these changes and provided some recommendations. It is wise to understand the directions of the change alongside these aspects and try to develop respective actions to sustain compatibility. Analyzing these aspects may also help support generating clear and implementable roadmaps. The changes and transformation of manufacturing systems, for example, are well characterized as shown in Figure 1.1. Similar analysis for other aspects listed above is given in the reference.

The digital transformation process requires special attention on some basic issues as they present challenges to the practitioners. It is important to consider these in any transformation process. Some of the challenges are reviewed below.

1.2.1 SMART FACTORIES

Creating intelligent operations in manufacturing functions to support the creation of so-called smart factories (unmanned factories or dark factories) should be the basic

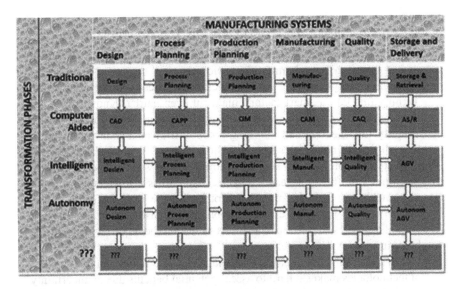

FIGURE 1.1   Transformation of manufacturing systems.

concern of practitioners. It is now very certain that the digital transformation cannot be complete without having intelligent and smart systems especially in manufacturing areas. Running unmanned systems will definitely require a great amount of knowledge and experience in related processes. This will in turn necessitate new skills and generate various supporting jobs. This is to say that the smart factories will in fact generate new jobs rather than leading to unemployment. Companies should be encouraged to spend time and effort to generate required autonomy as fast as possible.

So much is written about smart factories. This book will also provide some information; but taking humans out of the loop in a manufacturing environment and having a smooth-running system is not easy. This requires generating not only the machines, but also human behaviour that is able to cooperate with machines. With this understanding smart factories are considered to be the road to the digital factories of the future. See Grunov (2016) for more information. There are various challenges that have to be sorted out to run a smart factory. The challenges are twofold: in developing the smart systems and in running the smart factories.

Some of the challenges for developing smart systems are the following.

- Integration: Machines are not small equipment which can be changed easily. Keeping them running all the time without any problems is important. The digital designers should place a lot of effort in finding solutions to the problems without stopping the manufacturing lines.
- Connectivity: The machines in a smart factory are connected to each other and communicate one way or another. Sustaining this is important as the connection also serves as an information network carrying the information from one machine to others.

- Skills or behaviour: Machines are expected to perform the operator's behaviour. There are two issues here. One is having the skills to generate the required intelligence to the machines, the other is to model the behaviour of the machine on human-like performance.
- Fragmentation: Generating a smart machine requires extensive knowledge. Turning this into a smart factory makes it even harder. Knowledge from various scientific disciplines is required. Fragmentation is important to simplify the selection and planning process. It gives a chance for people from different knowledge sets to work together.
- Security: It is important to keep the information network isolated to prevent machines from being accessed and hacked. This is also essential to keep machines running in harmony with others in the manufacturing suites. Security issues may also create a problem with customers if trust is lost. Digital designers should make an effort to sustain the required level of trust by running the system safely.
- Uncertainty: When generating smart factories, the highly complex nature of technologies makes it hard to assess potential benefits due to uncertainty about the particular adaptation of processes and respective smart capabilities. The cost of setting up a smart factory is too high and the possible benefits are uncertain until realized. The investors should be made aware of the fact that the transformation is inevitable and be given support during the transformation.

Some of the challenges that are related to running the smart system as listed below.

- Process management: Manufacturing companies will always face difficulties in changing traditional routines and processes to those running digitally. As there is no systematically proven approach for this, there will always be a tendency to keep traditional practices. This rigid culture is difficult to change. The designers should spend a great deal of effort to generate applicable business transformation models both to enable the transformation and to attract the people with the competencies to support it.
- Supply chain: The ability to simplify a complex supply chain requires time and innovative ideas in order not to lose a quick and fast response to the market. Since global operations expand due to the impact of digital transformation, the supply chain must also extend to a growing scope of geographical areas.
- Reduced lead-times: Today, the business world is experiencing very fast delivery of products and services. It is therefore very important to reduce lead-times. Tailored inventory programmes are essential. This may help reduce development cost and enables fast lead-times for customer-specific devices. Achieving this still requires research and a disruptive technology.
- Product lifetime: Ensuring long-term availability of technology helps keep products in production for as long as possible. This, naturally, maximizes the investment value. Efficient management of failures and predictive maintenance contribute to continuity of the production line. Receiving customer

requirements and aligning the production line accordingly is another important aspect of sustaining product longevity. The designer should consider the functionality of the products as well as keep those in the market for longer times.

- Robustness and consolidation: This is as important as product longevity. Having high-quality and robust products, offering resistance to water, dust, impact, vibration and a range of other harsh conditions will be a key success to competition. This issue still remains open to new ideas.
- Demand management: Since smart factories allow fast response to customer requirements and customized products will seem to be available in the market, this may necessitate an active, efficient and effective demand management system. It would be wise enough to use this capability for marketing and the revenue generation process. In this respect the demand management system should have very specific capabilities which is yet to be developed.

Note that these and some other challenges not listed here introduce complexity to the transformation process. It would be nice if structured step-wise approaches to manage the large-scale organizational transformation of people, processes and technologies were developed and implemented. The framework proposed at the end of this chapter intends to provide this.

1.2.2 SWARM ROBOTICS

Swarm robotics is the study of how to design groups of distributed robots that operate without relying on any external infrastructure or on any form of centralized control. A collective behaviour (Garnier et al., 2005) emerges from the interactions between the robots and interactions of robots with the environment in which they are operating. Swarm intelligence and biological studies of insects, ants, bees and others are the main source of swarm behaviour. In swarm robotics, automatic design has been mostly performed using the evolutionary robotics approach (Nolfi and Floreano, 2000). Evolutionary robotics has been used to develop several collective behaviours including collective transport (Groß and Dorigo, 2008) and development of communication networks (Huaert et al., 2008). A swarm robot team is fault tolerant, scalable and flexible. The robots in a swarm environment are able to perform different activities concurrently. More importantly, swarm robotics promotes the development of systems that are able to cope well with the failure of one or more of their constituent robots. That is to say that the failure of an individual robots does not imply the failure of the whole swarm (Fault tolerance) as the swarm does not rely on any centralized control entity, leaders or any individual robot playing a predefined role.

Swarm robotics should also have a flexible structure where new capabilities can be embedded easily. The introduction or removal of an individual robot should not cause a drastic change in the overall performance of the system. Similarly, scalability is to be enabled by local sensing and communication. Each individual robot is to be responsible to interact with its environments using its own sensors and communication devices.

Without relying on pre-existing infrastructure or static information. The robots dynamically and automatically should allocate themselves to different tasks in order to match the requirements of the specific environment and operating conditions. However, introducing robustness, scalability and flexibility for solving real-life problems is still challenging. There are various factors such as unpredictable behaviours and uncertainties that prevent real-life uptake of swarm robots. Similarly, some effort is required on behavioural control to discover effective man-machine interactions, especially with a robot team.

In addition to those above, the digital transformation may also necessitate an engineering methodology which would cover the definition of standard metrics, performance assessment testbeds and formal analysis techniques to verify and guarantee the properties of swarm systems.

1.2.3 SMART AND SECURE NETWORKS

Digital transformation practitioners should spend some amount of time to understand the concept of smart network as it seems to be an important issue raised during the system changeover. A smart network is a collection of connected devices and machines that allows the transfer of data and gathers different kinds of information over business operations of an enterprise. With existing technology, all available data provided by both old and new devices may be presented to the dashboards for possible analysis. If this analysis automatically discovers certain trends and requirements to unleash new possibilities with the help of artificial intelligence and machine learning, this network is a smart network.

The smart network may make your data more accessible, meaningful and useful. As a consequence, it may allow identification of certain trends and patterns of the related industry. The growing trend to enrich manufacturing suites with smart technologies promises customer tailored products at low cost, at a high-quality and with a huge impact for companies, economies and societies across the globe.

Smart network systems may offer various services that include software services, security services, trainings facilitated by professional engineers. The success of digital transformation is directly related to the smartness of this network and it requires enormous transformative potential of smart technologies in all aspects of life. Being aware of this involves a range of challenges in the realms of technology, policy, society and business in order to stay competitive. Effective transformation requires a collaborative effort across any enterprise as a whole.

The integration of business systems, information technology and operational technology for the sake of data-driven decision making naturally poses new security challenges. This is important because newly connected systems may accelerate the speed and damage of attacks on communication networks. So far in business, many enterprises have hosted all of their communication and network devices on-site, especially when there are sensitive data. However, digital transformation urges enterprises to host their service on the cloud as explained above. This opens the connections to publicly accessible internet. Although there are some mechanisms that prevent and restrict data access, the digital technologies are still appetizing for potential intruders and hackers. Digital capabilities and other technological improvements increase the speed at which developers can create and update software systems making it more difficult to keep up with security processes and easier for security vulnerabilities.

On the other hand, new attack methods over the networks are continuously introduced. This necessitates the generation of new approaches and methodologies to cope with those. Note that, not only are the methods and methodologies becoming more sophisticated, but so too are the attackers. They can employ artificial intelligence or very advanced technologies (such as polymorphic attack) at the same level of security professionals and disable traditional security measures. For sustaining continuity of the transformation process, the network security must therefore be integral, agile, holistic and automated from the outset rather than brought together over time.

1.2.4 Simulation and Augmented Reality

Modelling operation environment through experiencing simulation and augmented reality (AR) should also be considered as key to success in digital transformation. In particular, the need for behaviour modelling is inevitable. Machines will perform the operations in the same manner and have to experience human-like behaviours. Once they are developed, they have to be tested in simulated environments under various scenarios to make sure that they are properly designed and respond properly to information acquired.

It is now certain that the future world will naturally integrate simulation and real life. Augmented reality, in this manner, will provide various opportunities to people, including supporting the use of very complex devices. This technology will also encourage generating machines capable of carrying out very complex operations through bridging reality and synthetic world. Furht (2011) reviewed this technology from various aspects.

Digital transformation has become very popular, especially after the introduction of augmented reality. However, this technology is still in its early stages and is evolving fast. Research in this vein includes use of head-mounted displays and virtual retinal displays for visualization purposes, and construction of controlled environments containing any number of sensors and actuators (Krishnaraj and Cherukuvada, 2018).

As Carmigniani and Furht (2011) state, the augmented reality systems allow people to interact with information directly without requiring the use of any intermediate device. They also point out that it should be very important for the developers to remember that AR aims to simplify the user's life by augmenting the user's senses, not interfering with them. Once successfully designed, augmented reality also brings the possibility of enhancing missing senses for disabled people. They can, for example, be used for hearing impaired users to receive visual cues, replacing missing audio signals.

Digital transformation practitioners who are planning to use this technology will face some challenges, such as social acceptance, privacy and ethical concerns. Some may even find it too costly to devise an AR system. Designers have to develop systems with easy integration capability. On the other hand, privacy concerns arise with the technologies that have the ability to identify and recognize people. There is a need for a system that allows people to decide which information about themselves is to be disseminated. Similarly, there are some ethical concerns as well. People hesitate to use the technology they think is science-fiction. Research results should encourage and aim to improve the level of use. This may support the implementation of a digital transformation process.

1.2.5   GLASS AS A VISION

An important component of augmented simulation is "vision" through glasses. Eyewear/headsets and eye-tracking devices are now becoming "intelligent" and eyes are directly serving as the connection to the internet and other connected devices. With direct access to internet applications and data through vision devices, an individual's experiences can be enhanced, mediated or completely augmented to provide different, immersive reality. Also, with emerging eye-tracking technologies, devices can feed information through visual interfaces, and human eyes can be the source for responding to perceived information. Enabling glasses as an immediate and direct interface can have an influence on learning, navigation, instruction and generating feedback for producing goods and services as well as experiencing entertainment.

This technology will also enable disabled people to engage more fully with the world. Once being equipped with this technology, immediate response to information will allow people and machines to take immediate actions and make informed decisions.

To utilize the glasses as a vision and to get the benefits of those in digital transformation still requires extensive research in order to face the challenges listed above.

1.2.6   BIG DATA AND INTERNET OF THINGS

Big data and Internet of Things should be well understood for the sake of fast and better transformation. This is extremely important and should not be ignored at all. More and more data will exist than ever before. Internet of Things (IoT) will be one of the key sources for generating this data. Since every activity is carried out mainly by machines through IoT and so much information flow is going to be possible, there is going to be a need to make the deep meaning of the data generated as explicit as possible.

Note that the data can have a different meaning in different times with different contexts. The ability to understand and manage this data will also have to improve. Once achieved, utilizing IoT and big data technologies together may allow people to make better decisions and be able to comprehend real impacts of the data and respective decisions. Keeping this in mind, the enterprises should invest in big data technologies to automate the data collection process and deliver new and innovative ideas to operators, even to the customers. This will definitely require automated decision support systems which can reduce complexities of the operations and services.

However, to utilize this technology, the trust in the data and decision algorithms will become the key issue. Since enterprises have a concern over privacy, establishing accountability and legal structures as well as clear guidelines for unanticipated consequences should be set up. They should work to promote awareness and opportunities of big data and IoT, together with some analytical tools and approaches. They should also develop means for the growth and development, capabilities for public-private data sharing, methods for impact analysis as well as possible transformation methodologies. Bhadani and Jothimani (2016) reviews basic challenges, opportunities and realities regarding big data. Rose et al. (2015) devoted their report on IoT especially to understand the issues and respective challenges.

It is possible to write so much about big data and IoT. The readers are referred to the references for technical information. This chapter draws attention to the main challenges that have to be faced. Practitioners of digital transformation should not ignore the fact that interpreting the data is much more important than collecting it. There is a huge amount of data available. They are hidden behind the machine operations, managerial decisions and people's interactions. Ingenuity is to derive meaning from these data and form a set of information for the decision makers. Note that the machines could also be the decision makers. Researcher and developers should make an effort to dig out effective business intelligence methods to help discover the real meaning of data in accordance with the current situation. Similarly, the IoT design process should also enforce some challenges such as communication, security, connectivity and integrity as discussed above for assuring smartness of the machines.

1.2.7   CLOUD NETWORK

Another important challenge in achieving digital transformation is to set up a cloud network and heavily utilize mobile systems. Since there is huge amount of data, there is a need to create effective data storage and retrieval systems. Cloud networking or cloud-based information network is a term describing the access of networking resources from a centralized third-party provider using Wide Area Networking (WAN) or internet-based access technologies. In cloud networking, the network and computing resources can be shared. This technology allows more network management functions to be managed by a cloud, so that fewer customer hardware or devices are needed to manage the network.

A standard cloud network makes it possible to create centralized data/information management, visibility and control of the overall network. This could include the management of distributed wireless routers or related office devices. The goal is to create and manage secure private networks by leveraging WAN connections and a centralized management function that can reside in a data centre.

Connectivity, security, management and control are pushed to the cloud and delivered as a service to those in need. As described by Rafael (2018), cloud computing consists of three distinct types of computing services delivered remotely to clients via the internet. Clients typically pay a monthly or annual service fee to gain access to systems that deliver software as a service, platforms as a service and infrastructure as a service to subscribers. Effective utilization of data and information during digital transformation requires a well-structured network which can be fully satisfied with cloud technology. The practitioners should seek a way to enrich their IT network with cloud capability for the efficiency of the transformation process.

Similarly, many organizations, particularly those with hybrid cloud environments report challenges related to getting their public cloud and on-premise tools (legacy systems) and applications to work together. Digital designers should be aware of the fact that integrating legacy systems and cloud-based applications requires time, skill and resources.

In addition to these, there are also other challenges involved in cloud operations. They are:

- Cost: Cloud computing itself is affordable but tuning the platform according to the company's needs can be expensive due to additional bandwidth and continuous scalability. The expense of transferring the data to public clouds may be a problem for short-lived and small-scale projects.
- Reliability of service providers: The technical capability of service provider is also important. It should provide continuous (seven days a week/24 hours a day) support and keep the system up and running. The techniques via which a provider observes its services and defends dependability claims should be well comprehended.
- Downtime: This is a significant problem in cloud technology. Downtime is unavoidable. A trustworthy internet connection has to be set up before adopting enterprise operations to cloud networks.
- Security: The more people that access the cloud the less secure it is. Employees should have strict control over their passwords. The designers should employ a multi-factor authentication system to sure that the passwords are protected and altered regularly, particularly for those who leave.
- Data privacy: Enterprise-specific information should be kept for internal use only and not be shared with third parties. Digital designers should have a plan to securely and efficiently manage the data they gather.
- Expertise: While many IT workers have been taking steps to improve their cloud computing expertise, employers continue to find it difficult to find required IT specialists. It is obvious that the technology workers with knowledge of the latest developments in cloud, open source, mobile, big data, security and other related technologies are on course to become more valuable in the following years. The digital transformation process will definitely require talented experts of computing.
- Compliance to regulations: Many nations are enforcing some regulations on cloud computing. For example, they may require a data protection expert who is responsible for overseeing data privacy and security. Enterprises should try to meet any legal or statutory obligations. This also puts some burden on the digital designers to overcome.
- Migration: Although a straightforward process is defined for launching a new application in the cloud, moving an existing application to the cloud is still not easy. Many enterprises find it time-consuming especially in terms of troubleshooting, security configuration, data migration and synchronization. IT specialists should perform pre-migrations tests over and over again to make sure that the system will be operational after migration.

- Immature technology: Many cloud computing services are on the cutting edge of technologies such as artificial intelligence, machine learning, augmented reality, virtual reality and advanced big data analytics. However, the enterprise expectations in terms of performance, usability and reliability are not totally fulfilled. The main problem for this is the use of immature technology. Digital transformation practitioners should make sure that their requirements are fulfilled. Otherwise they have to find alternative solutions.

1.2.8 MOBILE SYSTEMS

Similar to cloud computing, mobile systems which also need certain attention are also the essential components of future systems. They can be considered as the set of IT technologies, products, services and operational procedures that enable the users to gain access to data, information and related capabilities over mobile systems such as mobile phones.

Note that an important feature of these systems is that they are unrestricted to a given geographic location. The applications of mobile computing systems are becoming ubiquitous and pervasive in the business, consumer, industrial, entertainment sectors, etc. as they allow users anytime, anywhere access to information and computational resources. With these capabilities "mobility" is often regarded as an integral part of the transformation process. Enterprises will be heavily engaged in using mobile devices and mobile applications to solve their business problems.

Similar to cloud computing there are some challenges ahead for practitioners of digital transformation regarding mobile systems. Wireless communication has different characteristics than wireline communication. The bandwidth, latency, variability and reliability of the line should be designed in accordance with mobile standards. Highly adaptive applications that can even support disconnected operations, with a possible level of autonomy, are required. Because of mobility, resource discovery, privacy and the migration of locality become important issues. On the other hand, the digital world is heavily employing portable computers. They face physical challenge, pragmatic challenges and systems issues. The transformation process should not ignore these facts before kick-off.

1.2.9 ADDITIVE MANUFACTURING AND 3D PRINTING

Digital transformation is also facilitated with new approaches such as additive manufacturing, often referred to as 3D printing. This technology incorporates a wide variety of processes and materials that share a common characteristic which is the direct transformation of 3D data into the physical objects. One of the main expectations of digital transformation is to speed up time to market, especially when there are changes in tools and fixtures as well as production lots and batches.

3D printing prevents the use of cumbersome tools with various revisions. This naturally eliminates the time and cost associated with production changeover times. It is simply transferring a digital CAD file to an object printer and generating a physical copy of it. With this capability, additive manufacturing significantly reduces the design cycle. Another benefit of this technology is that virtual teams across the globe can collaborate for a shared design activity by utilizing CAD files.

Someone in the United States can design a fixture and determine the best practices. In a very short time (say less than 24 hours) someone else in Turkey can take these practices and implement them into his work without spending too much time reinventing the wheel. This capability enables generation of global work standards as well. The digital transformation process seems to get the most benefit out of this technology. Additive manufacturing provides various opportunities for different sectors. See, for example, Carlström et al. (2015) for possible applications in the metal industry. This report provides a basic introduction to additive manufacturing and makes some recommendations for designers and engineers who deal with 3D printing.

1.2.10 CYBER-PHYSICAL SYSTEM

Once introduced, cyber-physical systems (CPSs) are at the top of the agenda in any digital transformation process. A CPS is a mechanism that is controlled or monitored by computer-based algorithms, tightly integrated with the internet and its users. It is an integration of computation, networking and physical processes. That is to say that embedded computers and network devices monitor and control the physical processes providing abstractions and modelling, design and analysis techniques for the integrated overall systems.

A CPS mainly can be considered as a feedback system that is distributed and networked over the whole enterprise, possibly with wireless sensing and actuation. It is adaptive, intelligent and predictive which can work in real time. It requires improved design and design tools as well as a proper design methodology that can support specifications, modelling and analysis of networking, interoperability and time synchronization. It is also required that a CPS supports scalability and complexity management through modularity, composability, synthesis and interfacing with legacy systems.

Validation and verification of simulation models, stochastics models and assurance systems can also be supported by CPSs. Similarly, cyber security such as intrusion detection and cyber-attacks is also the main concern. Today CPSs can be applied in many domains including communication, manufacturing, military, robotics, smart buildings, transportation, physical security and infrastructure, etc.

The practical implementations of the paradigm of cyber-physical systems appear in many different forms. However, they can be identified based on their distinctive characteristics such as distributed, multi-scaled, dynamic, smart, cooperative and adaptive. Horváth and Gerritsen (2012) defined the following characteristics which require attention in the development process.

- CPSs are designed and implemented to support human activities.
- CPSs are functionally and structurally open systems with blurred overall system boundaries.
- CPSs have the capability to change their boundaries and behaviour dynamically.

- CPSs consist of a digital cyber-part and an analogue physical part.
- CPSs are articulated and heterogeneous and are constructed of a very diverse set of components.
- CPSs manifest on various extreme spatial scales and temporal ranges.
- CPSs can provide real-time information processing capability.
- CPS components have either predefined or ad-hoc functional connections.
- CPS components are knowledge-intensive and able to handle both built-in knowledge and the knowledge generated by reasoning and machine learning.
- CPS components are able to memorize and learn from history and situations in an unsupervised manner based on smart software agents.
- CPS components are able to execute non-planned functional interactions.
- Overall decision making is distributed over a large number of components based on reflexive interactions.
- In order to maintain security, integrity and reliability of the components, different sophisticated strategies are implemented.
- Next generation CPSs (molecular and bio-computing based) are supposed to have some level of reproductive intelligence.

When designing a digital system most of these characteristics should be taken into account. Each of these may require a specific type of knowledge and extensive experience which makes the development of CPSs a real challenge.

1.2.11 Preventive Maintenance

The practitioners working to achieve digital transformations should aim to sustain quality through preventive maintenance. This is also a challenging objective. Preventive maintenance presents systematic inspections and related processes to identify and correct potential problems before they occur. This definitely increases the lifetime of equipment and improves reliability of machine suites in addition to sustaining employee safety. This, in turn, minimizes repair costs and production slowdowns as traditionally, a significant part of resources for maintenance are wasted on inefficient use. Digital transformation presents a big opportunity for maintenance systems.

It is now well expected that the digital transformation in maintenance systems facilitates reduction of unscheduled downtime by avoiding equipment failures and shortening the duration of the turnaround. At the same time, the maintenance cost is expected to be reduced.

On the other hand, smart equipment monitoring systems can improve plant maintainability and help with new data-driven reliability programmes. The digital transformation of how manufacturing suites are run and maintained should be an important concern for transformation practitioners.

Monitoring the conditions of machines, generating predictive maintenance plans and environment in which the plans are generated, identifying and simplifying maintenance operations, validating PM plans through simulations and augmented reality are some of the challenging activities. Basri et al. (2017) provided an extensive review of PM systems and related implementations.

1.2.12 GREEN TECHNOLOGY

The digital transformation urges "green manufacturing" which is the renewal of production processes and establishment of environmentally friendly operations within the manufacturing area. Digital designers encourage this by either producing "green products", particularly those used in renewable energy and clean technology equipment or implementing "green manufacturing" for reducing pollution and waste through recycling and reuse of waste. While performing green manufacturing it is important to take regulatory compliance and pollution control into account. Green manufacturing is also known by different names, such as clean manufacturing, environmentally conscious manufacturing, environmentally benign manufacturing, environmentally responsible manufacturing and sustainable manufacturing (Rehvan and Shrivastava, 2013).

Although digital transformation does not explicitly refer to ecological sustainability of production systems as a major step ahead, the production technology and operations research community has addressed ecological impact and sustainability in social progress. It is assumed that only the production systems that incorporate ecological sustainability in their concept of intelligence will be competitive. The main concern of transformation is, therefore, to endure hidden, long-term costs caused by poor environmental behaviours. Keeping this in mind, enterprises across many industries have made advances in reducing single use plastic, carbon footprints and water waste.

While manufacturing requires energy, it doesn't require waste or necessarily involve traditional fossil fuels. It is reported that, by innovating at multiple steps in the manufacturing process, reducing energy use by 3–5% at each stage, overall "dirty" energy could be reduced by 30–60% (Sealevel, 2019). Doing research on energy, fuel-cells, solar energy use, energy security and lightweight materials may help companies to utilize the technology achieved to reduce their overall carbon impact.

Not only the plastic packaging materials – even non-plastic ones – often waste water during their creation process. Packaging materials of steel, aluminium, glass, paper and plastics can waste up to 800 billion cubic metres of water annually (Sealevel, 2019). This waste comes in the form of non-reusable water and polluted, restorable water. Since plastic packaging will always have a place in certain manufacturing spaces such as pharmaceutical and medical operations, digital transformation requires that the practitioners utilize advanced packaging science to reduce material consumption overall and diminish packaging's environmental impact. Technologies such as automation and flexible electronics have led to plastic production systems that minimize water use and lead to immediate wastewater cleaning, drastically reducing the impact of plastics on water systems.

1.2.13 SMART CITY AND RELATED APPLICATIONS

The term "smart city" was used for pointing out the implementation of user-friendly information and communication technologies developed by major industries for urban spaces aiming to provide high-quality lives for the residents of a city. This concept promotes social and technological innovations and links existing infrastructures to technological progress. Note that smart cities not only incorporate new energy, traffic and transport concepts making life easy for people, but also focus their attention on new methods for governance and urge public participation. Achieving these may differ from city to city depending upon the regional trends that illustrate that there are divergent urban growth patterns among major regions with different levels of economic development. There are three areas of interest in making the cities as smart as possible. Digital opportunities are utilized to support these and respective sustainability. They are economic, social and environmental sustainability (Breining et al., 2014).

Smart cities should also tackle the current global challenges, such as climate change and scarcity of resources. Digital transformation practices are expected to enable optimum utilization of the resources and provide alternative approaches to deal with unexpected natural phenomena. Similarly, securing economic competitiveness and quality of life for urban populations continuously is also a rising concern.

The smart city development process should progress in two ways: technological infrastructure and social welfare. These two should be paired together in considering the digital transformation. That is to say that the technology-oriented approaches should be implemented without ignoring welfare and social aspects. New technologies must be assessed as to their benefit for the public interest and the preservation of creative freedom in public spaces.

Smartness of cities should be enriched by implementable and publicly acceptable smart ideas. Accessibility, affordability and safety of city systems, as well as compact urban development are essential factors in this context. Digital skills must be acquired to handle the new tools with care, especially with regards to big data applications and data security, energy, mobility, economy, society, politics, administration for the sake of public life quality. Technical, economic and social innovations provide the foundation for these activities and promote sustainability. Similarly, smartness can bring more diversity of operations. Having a digitally managed and shared economy, for example, has garnered a great deal of interest in recent years. Joint utilization of private facilities instead of private ownership is likely to continue to be at the top of the agenda for city developers.

A smart city project team leader and his team members (Breining et al., 2014) clearly stated in their white paper that cities are facing a number of environmental sustainability challenges, generated by the city itself or caused by weather or geological events. Digital transformation promotes efficient and intelligent deployment of technology and aims to integrate infrastructures in order to reduce the impact of the city on the environment resource. This generates smart technologies which bring about value to the city. They also emphasized that the digital transformation of the society, implying smartness, helps cities to improve efficiency, enhance their economic potential, reduce costs, open the door to new business and services, and improve the living conditions of its residents. Furthermore, they explained that most of the smart city applications mainly focus on vertical integration within existing independent infrastructure and services silos, e.g. energy, transport, water or health. However, real "smart" city implementations require horizontal integration to launch systems capable of generating new opportunities for the city and achieving considerable increases in efficiency of the services. Note that electric grids, gas/heat/

water distribution systems, public and private transportation systems and commercial buildings/hospitals/homes play a key role in shaping sustainability of the cities. Thus, interoperability of the city systems (horizontal and vertical integration) is essential.

Since the technology utilization in a city will have to include vertical and horizontal integration as stated above, the data from the different sources and services will need to be combined for preventing risks and managing the future cities. This integration is/will therefore be the main concern of digital transformation; the city planning may otherwise lead to unexpected inefficiencies with suboptimal outcomes and higher costs.

There is a wide variety of decision makers and decisions to be carried out in a city. Social and technological transformation, therefore, cannot be easily encountered. With the new shape of the world, the citizens or so-called stakeholders (leaders, service providers, managers, operators, end users, investors, religious congregations, academia, charities, etc.) are not only considered to be the users of the services but also the "service providers" in accordance with their expertise. It is now very well known that smart services and infrastructures in a city cannot be developed without proper collaboration of those. Exchange of fundamental data on customers, infrastructures and operations is a must and the lack of this is one of the most important barriers.

Digital designers should take this into account when they generate "city plans". Stakeholders should be satisfied, and their concerns need to be carefully traced to prevent obstacles before possible changes and follow the managerial vision.

There is still a great need to develop smart city standards to provide a general framework for assuring data exchange and developing the required smartness through IoT, CPS and cyber security. Since there are plenty of system developers and software houses working in parallel to create the tools, systems and products that provide smart enough solutions, internationally agreed standards that include technical specifications to support required interoperability are needed.

Besides, the standards should be easy to understand especially by non-specialists such as city managers as well as mid-level managers to incorporate those into operations. The challenges stated above obviously indicates the need for extensive research on digitization of cities. Maheswaran and Badidi (2018) provide information regarding several aspects of smart cities.

1.3 FACTS ABOUT THE DIGITAL TRANSFORMATION

In order to have a clear understanding of digital transformation, practitioners (managers and operators as well as designers) should be aware of some of the facts about digital transformation. Some of them are explained below.

Fact 1: "Human" is still the main driver of the digital transformation process and "behaviour modelling" is a must.

When talking about digital transformation, unmanned systems and autonomy is considered to be the key issue for success. However, generating required autonomy still necessitates human intelligence. While automation requires machine modelling which can work with electric power, the autonomy is more beyond that. It requires both machine modelling and behaviour modelling. Human intelligence and knowledge processing capability are to be transferred to the machines.

This can only be done by the human intervention or the directions produced by the human intelligence. This clearly indicates that the human being will be the most important and unavoidable actor of the transformation process. This implies another fact of this process that is knowledge utilization capability.

Fact 2: Energy of digital transformation is the "knowledge".

Generating autonomy behaviour of systems, digital capabilities and functionalities requires utilization of an extensive set of domain knowledge. Artificial intelligence, robotics and sensors and information technologies are brought together for digitizing the systems. Innovations through knowledge-intensive approaches are an unavoidable requirement for the continuity of the transformation process. Without prior domain knowledge and proper knowledge flow channels, this continuity cannot be sustained.

Setting up respective hardware and infrastructure without knowledge and knowledge-processing capabilities will not contribute enough to the digitization process.

"Knowledge" should be considered as the energy of the future systems. The importance of knowledge should be kept at the top of the agenda by the practitioners of digital transformation.

Fact 3: Technology is definitely not the starting point.

Digital transformation is emerging by technological improvements and transforming the systems to those digitally operated. However, the technology is definitely not the starting point. New technology in an old organization creates an expensive old organization. It does not produce change.

The most essential baseline and unavoidable requirement of transformation is leadership and strategy. Giving up traditional approaches and working habits is not easy, especially if no operational problems exist. There is a proverb "no need to bring new traditions to an old village". This applies to all enterprises. Everyone should understand the need and should develop an intention to follow.

The decision makers should develop a "communication plan" and an "awareness plan" before starting the transformation process. The leadership includes defining goals, articulating strategy and empowering the workforce to reach its highest potential to own and implement the changes. Change leaders are therefore the driving forces and momentum that enable large-scale transformation.

Fact 4: Ownership of digital transformation is essential for success.

Before starting any transformation process, the people of the organizations, including those in government agencies, should believe the need for and possible benefits of digital transformation and they should own the process. This is a process that has to run in two ways: managerial and operational efforts. No transformation will possible if it is left only to the operational people to change the systems they are operating.

Managers and key stakeholders should understand the power of change and basic drivers. Before starting the process all key stakeholders should come together and agree upon the need and the process to follow. Generating a change climate and promoting this throughout the organization will allow operational people to be more encouraged in implementing the change wherever necessary. This implies the capability of managing the technology for better and valuable use.

Fact 5: If technology is not managed, it manages the followers.

Technological progress is unavoidable. Technology presents its value and benefit to those who can manage it. There is no way that the technology stops and waits for others to come and catch up. If enterprises manage the technology right at the outset, it will be possible to generate guidelines for the technological progress to follow for the sake of better change management and value generation. Otherwise others who manage the technology will define the rules of the game and the technology itself will manage the enterprises or followers. This may be a big barrier in digital transformation and will also prevent achieving possible values and business benefits out of technology. Technology management capability should be enriched by estimating the progress along business and social progress.

Fact 6: Technologic forecasting is essential to comprehend the future systems.

Things are changing. In the past, it would not be possible to comprehend current systems of today using the systems of those times. Similarly, present systems and approaches would not be enough to estimate the future. The trends and possible future implications should be taken as the baseline. That is to say that system designers for digital transformation practitioners should be able to read the future with "future glasses". If this is achieved, there will not, for example, be a fear of losing jobs as the transformation will bring about so many new professions which do not yet exist. This also imposes another fact of digital transformation. People will not be engaged in similar activities in the future. They will have to develop new competencies to cope with the current business requirements.

Fact 7: Digital transformation requires new sets of skills.

Digital transformation may not be successful in utilizing only the existing competencies. It imposes a new way of doing things and leaves operational burdens mainly on machines. The people are expected to utilize their intellectual capacities rather than physical powers.

It is now very obvious that new professions will emerge and some of the existing ones will disappear. There will be a strong need for those with digital talents and skills. The ability to quickly adapt the system to the digital world and the change will be the prerequisite for holding a job. It seems that the skills such as agility, adjustment and adaptation to the new opportunities will be more important than technology skills. However, one of the important issues in this respect is that the enterprises, while seeking digitally competent employees, should not lose their existing competencies. During the transformation process, both are needed.

Fact 8: Digital transformation is not a one step effort. It requires time and heavy analytical thinking.

The transformation process should be carefully implemented throughout a time span. It will consume time and require much analytical thinking. The practitioners should build their own ways for success and learn as they go along by adapting their real-life experience to changing external market conditions. Managers and designers should analyze the growth, new expansions, sectoral developments with the possible impact of digital disruption. They have to understand the capability of their organization (resources available, obstacles, financial availability, cultural impacts, etc.) to embed the transformation. They also have to define the priorities and immediate actions necessary before starting the project.

Fact 9: Transformation success strictly depends on multiple views.

The transformation process is having a direct effect on employees and customers. Customer needs are, in fact, driving the process. Employees should be ready to perform operations in a new manner with limited intervention. This needs to be taken into account when making a change in any operation or process of the enterprise. However, this is not enough. As outlined by Gursev (2019) this process requires specific attention on the following viewpoints.

- Strategic view: This includes supporting the transformation process in terms of assuring cultural transformation, developing a strategic roadmap, performing technology watch, developing the leadership required and sustaining innovation management, etc.
- Managerial view: This includes the transformation process being shaped around business, operations and product. Generating an action plan on process digitization (mainly implementing CPS), utilizing the information technology (mainly IoT and cloud) and security, orchestration of services, emergency handling, modelling and simulation, personalization of the processes and sustaining the required level autonomy, etc. are some of the main concerns of this view.
- Technical view: This view deals with the technological background of the transformation. It includes artificial intelligence applications for smartness (smart network, smart factory, smart sensors, smart city, etc.), machine to machine performance, equipment infrastructure, data utilization, 3D printing, mobile devices, information system applications (manufacturing execution systems, Enterprise Resource Management, multilevel customer interactions and Customer Relation Management systems), cloud computing, new generation technologies (nano and bio technologies) as well as technological standards. Effective and efficient utilization of these are essential for success.
- Human resource view: This deals with people (employee skills, competencies, openness to new ideas, autonomy required), regulations (audits, legal restrictions, labour regulations, protection of intellectual property), training (on-the-job, programmed) and man-machine interfaces (user centric design, user interface design, value sensitive design).

These different views are to be taken seriously when the organization develops transformation strategies, generates respective implementation plans and carries out technical innovations and infrastructure modifications with competent labour forces.

Fact 10: The importance of communication should not be underestimated.

Most of the digital transformation efforts fail, not due to a fundamentally flawed technology and implementation roadmap, but due to communication failures. Collaboration and communication at every stage of transformational change, understanding of why things are changing, how they are changing, what impacts they will have and what the possible the benefits are, should remain at the top of the agenda.

Fact 11: Long-life learning systems within enterprises will be the main platform of knowledge dissemination.

The digital transformation process is a never-ending story. The speed of knowledge generation is so fast. There is always a need for a platform to share a new knowledge set with respective implications. This can only be sustained through on-line learning platforms. Each enterprise will have to set up the means to play the role of training institution within the requirements of the transformation and the needs of the enterprise. Company-specific knowledge dissemination engines would help increase the awareness as well as support the need to accomplish required capabilities. On-line learning systems will also have to be enriched by real-life cases, prototype implementations, on-line discussion sessions and employing the systems for reaching available knowledge outside the company. Customized systems look as if they will be prominent in the near future.

Fact 12: Understanding the main drivers of the digital transformation is essential for a successful progress.

The digital transformation process should not be considered as usual progress in society. There are some driving forces. The practitioners should spend a great amount of time understanding these drivers and perform digital operations accordingly. The transformation strategy, action plans and running process should be generated by taking these into account. Some of them are listed below.

- Innovation: That is, imposing new ideas into product generation and process designs. Even social innovation is driving the changes.
- Customer expectations: New systems and functionalities generate new expectations of the customers. To sustain competitive advantage, the products and services should reach beyond customer expectations and provide more functionalities. There are always better opportunities in the digital world to provide this.
- Rapid response: In addition to new features and functions, the digital operations improve and increase the speed of operations. Most of the operations are carried out automatically, some of them even with a degree of autonomy. Being able to respond demand and requirements as fast as possible may help sustain competitive advantage. This is therefore one of the factors that triggers the digital transformation.
- Enhanced security: Utilizing information technology naturally brings about information security problems. Existing IT networks of enterprises may not be capable enough to deal with cyber security requirements. This may force them to renew their infrastructure to be able to cope with technological threats. Keeping information private is also the main issue. This definitely

requires the renovation of knowledge channels. There is a mutual impact in digitization. On one hand, digital transformation may lead to secure information networks through smart networking. On the other hand, the smart networks may encourage digital transformation.

- Digital marketing and consumption: Digital transformation provides various opportunities for the products and services to be marketed. Social media is now a good source for commercialization. The enterprises may have a chance to reach direct customers faster than ever before. They do not need to spend so much effort to identify the right set of potential customers. The transformation process should benefit with digital marketing capabilities which, in turn, triggers more digitization.
- New business models: The digital world offers new business models. Some of those are subscription models (such as Netflix, Apple Music) where users continue to access the products and services without really buying it; freemium model (such as LinkedIn, Dropbox) where users pay for a basic service or product with their data or "eyeballs", rather than money; marketplace model (such as eBay, App Store) that brings together buyers and sellers directly; ownership model (such as Zipcar, Airbnb) which takes a commission from people to monetize their assets (home, car, capital) by lending them to "borrowers"; experience model (such as Apple, Tesla) which provides a superior experience, for which people are prepared to pay; on-demand model (such as Uber) which takes a commission from people who use and pay for goods and services delivered; and hypermarket model (such as Amazon and Apple) which use market power and scale to crush competition, often by selling below cost price. To be able to deal with new business models, digital designers should develop internal regulations as well as reshape enterprise strategy for involving new processes. They can use property rights and copyright infringement to protect the new models. They may spend some efforts to receive the most possible value from vulnerable business and they can choose to invest in disruptive technologies, human capabilities, digitized processes or perhaps acquiring companies having these attributes. Alternatively, they can introduce a new product or service to the market aiming to develop competitive advantage on size, market knowledge, brand, access to capital and relationships to build the new business. They can even define a new business model in an adjacent industry to sustain competitive power or they can think of completely changing the business to new areas. It seems that seeking new business models will remain one of the main drivers of digital transformation.
- Global competitiveness: The world is getting smaller with digital operations. Companies are finding opportunities to conduct business all around the world. Being completive in national markets is not enough to sustain market share anymore. There is a need to develop international relations and global business with some competencies that will take the enterprise to the fore.
- Digital assistants: The digital world does not only support manufacturing operations. It also provides some opportunities to develop digital assistants

to make life easier and of higher quality. They attract the attention of entrepreneurs due to their multi-use capability. It is well known that they can support their users by providing timely information and knowledge which would, in turn, lead to better business results.

- Expensive labour costs: This is another driver that triggers the digital transformation. Being able to perform operations with minimum human intervention naturally reduces labour costs. Digital operations may therefore attract the attention of entrepreneurs as they seek every opportunity to reduce the cost to sustain cost-sensitive competitiveness.

Fact 13: Implementing the new technologies is a must.

To be able to create fully digitized systems, the digital transformation process should continuously be fed by new technological achievements. Some those are already mentioned above. Emerging technologies for enabling digital transformation is not limited but includes robotics, artificial intelligence, cloud computing, big data and data analytics, Internet of Things, 3D printing and Blockchain. Note that the digital transformation will not be complete without utilization of these one way or another. Note that these technologies are not new, and they have been utilized for so many years now. It is important to embed the right digital technologies into operational systems. Similarly, the ecosystem should be alerted and aligned accordingly.

Fact 14: Digital transformation impacts not only the industry but also have a direct influence on the society as a whole.

Digital system designers are generating products and service which directly lead to the transformation of the society. In this regard, digital transformation should be considered as a continuous stream of change. The changes in all areas such as manufacturing, health, trade, finance, transportation, energy, public administration, etc. trigger digital systems to emerge whereas the progress on digital systems triggers the societal changes. This implies that everyone should make every effort to realize digitization for the overall nation-wide digital transformation.

Fact 15: Enterprises must be open to radical reinvention to find new, significant and sustainable sources of revenue.

Enterprises should always look for opportunities such as new business models. They must seek new innovative ideas and must be prepared to tear themselves away from routine thinking and behaviour. This is more important than spending money and making investments for digital transformation. This obviously generates some risk due to uncertainties. However, it is always worth trying to find and implement new ideas in order to improve effectiveness of products, services as well as processes.

Fact 16: Digital transformation cannot be imported from outside.

The people within the enterprises including all stakeholders should spend a great amount of time and effort to achieve a sustainable transformation and continuously update the process in light of technological progress. They can have some support and consultancy, but they cannot leave this process to be operated by consultants. It should be well understood that this process is expensive in terms of time and other resources. Employees must understand the dynamics of the transformation very well and manage that by themselves through continuous improvement. The real fact is that those looking outwards for a solution will always pursue the lack of solution.

Fact 17: Everyone starts the process from the same point, but the speed is too high.

The transformation process outlines some of the underlying technologies to be digitized. However, this is necessary in every aspect of life and in every enterprise doing business. All competitors are running transformation projects at the same time. The more concentration that is given, the better the success achieved. Whoever adopts more systematic approaches and a clear roadmap will be further ahead. The speed of change is too high. If not followed properly, it will be hard to keep up. For this reason, it is necessary to manage the change process in the whole enterprise and continue to transform in the direction of change. Otherwise it will not be easy to close the gap. Changes and industrial transformations in the past provide a good hint for the speed of the current stream of digitization. Note that it took millions of years to progress from a hunting society to an agricultural one. It then took tens of thousands of years to reach the first industrial revolution. After about 300–400 years, the second industrial revolution appeared and lasted about 150–200 years. After the introduction of the third revolution it took about 50–60 years for the fourth one to emerge. Now the digital transformation is expected to set a different pace and take a different shape in the coming 20–30 years.

Fact 18: Digital transformation starts with the "first" and the "right" initiative.

So much is written about digital transformation and industrial revolutions. There have been so many conferences, panel discussions, seminars and workshops. These activities encourage the practitioners but are not enough to start with. They look for a possible action with which to start the process. Some spend months understanding what is going on around them. Some makes move with little efforts and small changes. Having the first implementation from the right starting point assures both quality and continuity of the process. System designers are urged to develop a clear roadmap as outlined below and take the first initiative.

1.4 HOW TO COPE WITH THE TRANSFORMATION

As clearly stated above there are a huge amount of challenges that have to be dealt with in ensuring digital transformation. Managerial authorities and digital designers should follow a systematic approach to sustain the required level of digitization. Governments should create a clear roadmap for social transformation based on digital systems. This section provides some information on how to achieve this.

1.4.1 SETTING UP A CLEAR VISION AND A STRATEGY

This is the most important step in an effective digitization process. Vision and strategy, by their definition, imply the true intent of the management and are the baseline for operational objectives and targets. They have to be well understood, comprehended and internalized. Similar to all other changes, the digital transformation process should be fed by those and the roadmap generated should rely upon those. It is now well known to the industrial community that not so many projects are success-ful and assure full digitization.

It is essential, in particular, for the decision makers to have a clear understand-ing of the starting point and the direction to move ahead. Instead of dealing with complex associations and trying to solve internal crises by urgent efforts, a focused methodology should be implemented to deliver the transformational changes through consistency of purpose. In order to achieve this, creating a vision and strategy is an essential part of this process.

Digital transformation will not be successful without the necessary ownership. After this, developing a clear implementation strategy and experimenting opera-tional excellence will lead to digitally led success and will bring competitive advan-tage, disruption avoidance, cost reduction and revenue growth.

The strategy to achieve transformation will not be complete without including priorities, ambition and outcomes, along with relevant performance indicators as well as possible risks of the investments.

Strategy should direct and focus the attention of all stakeholders on all kinds of actions, behaviour, processes and interactions (internally and externally) that should work in harmony to drive the business forward effectively and efficiently to its desired transformation.

Vision and strategy should play the role of an umbrella for developing corporate regulations, developing new business models, data sharing platforms and e-opera-tions. This has to be assured by a systematic strategic management framework with several steps such as:

- Planning: This is to develop an agile, iterative and controlled process to define the strategy. This step should also include the process of developing a comprehensive insight into the technological developments in the sector in which the enterprise is operating. Before making the plans, existing com-pany strategy, known lessons, risks and issues should also be appraised.
- Development: This is to generate a crystal-clear direction of the enterprise under the available resources, priorities and both intellectual and physical capacities. This process should define objectives and targets for possible digitization to achieve. Establishing the best practices across all digital markets, delivering digital product development roadmaps, visualizing processes, embedding new technologies into processes, etc. should be directed by this strategy.
- Communication: That is to make all stakeholders aware of the new strat-egy. Communication channels should be open at all times for any kind of data and information exchange. Management should show strict ownership of the process and not allow cynicism around digital transformation and change to nullify the efforts. It is also necessary to maintain regular loops of feedback during this process and keep everyone updated and engaged.
- Execution: This step is finding out an optimum way of implementing and monitoring the execution of the predefined strategy. To sustain this capa-bility an "implementation plan" should be prepared and implemented. By monitoring implementation regularly, it may be possible to make necessary adjustments to the strategy and update the plans in time.

In the near future, developing a vision and a strategy aligned with the change speed and respective content is expected to: generate a good atmosphere for those employed, help understanding of how digital transformation can upend business models, focus attention on opportunities in the future, remove barriers and create incentives, engage change story and rationalize, encourage and establish better innovation and competitiveness.

1.4.2   GENERATING AND IMPLEMENTING A TRAINING PROGRAMME

As mentioned above, digital transformation will become more powerful and effective if there is participation and ownership. This can be facilitated to a greater extent by an awareness programme. Education and training are the baseline for this. All stakeholders of the enterprise, even the customers, should be involved in one way or another in the training programme. If the key actors do not have enough educational background, the first thing is to get well equipped with missing qualifications. Collaboration with educational institutions may help this in order to overcome inadequacy.

Since existing knowledge is always improving and technological progress is bringing about new knowledge, it is essential to convey a new set of knowledge to key actors of the transformation processes. This implies generation of and applying a continuous training programme which would at least involve the following issues.

- Understanding digital transformation and its social impacts.
- Main drivers of this transformation.
- New technologies emerging such as digital manufacturing, artificial intelligence, robotics and autonomy of those, integrated communication network (RFID and IoT), cyber-physical systems, smart factories (flexibility, speed, productivity), big data and analytics, business intelligence.
- Innovation methods and methodologies and respective processes.
- Change management processes.
- Core competencies of the enterprise with respect to digital transformation.
- Intelligent products and processes.
- Smart systems and respective utilization.

Note that this training programme should be employed on the "basis of the need". Not all should enrol on the whole programme. In the digital world, training to fill skill gaps should be offered on-line and on a just-in-time basis due to the speed of change. Training platforms with the capability of offering videos, mobile and digital content to employees will be necessary.

1.4.3   CREATING AN ACTION PLAN TO IMPLEMENT THE STRATEGY

Creating an action plan is as important as generating strategy. It is not possible to generate and implement a standard action plan. It changes from business to business, from enterprise to enterprise. Action plans should be short-term, mid-term or long-term, each with a different motivation behind it. Although there is no standard plan to implement, the following will characterize the respective actions.

- Generating a framework programme for digital transformation: Each company should define a generic framework for digital transformation. This can be considered as a main platform for running the digital transformation process. This framework programme may play the role of a guideline to direct the activities for transforming enterprise systems. It should present a long-term implementation plan for transforming the enterprise step by step through pointing out the objectives, scope and respective standards. It should also be shaped in accordance with available company resources. It will also be possible to systematically follow the progress under this framework and receive expert views upon request. The transformation framework should be company specific and provide a foundation on which operational process is to be digitized and in what way. The framework may also allow reuse of some of the digital components through establishing software code libraries, generic simulation systems and digitally ready to use operational specifications. It should support the digitization process in four steps as given below.
- Generating cultural atmosphere and awareness of digitization: The main aim will be supporting digital literacy with training and education and developing digital mindset. This may include training programmes, workshops, company visits, on-the-job experiments, competition on how to get digitized, tea and talk seminars, etc. Continuous learning on innovation and technologic motivation should also be the main concern.
- Setting up company-wide standards for developing digital transformation: This effort should aim to make the following as clear as possible.
  - Basic expectations of digitalization and SMART objectives set around the strategic objective.
  - Review of the organization's digital capabilities.
  - Conducting benchmarking with competitors wherever possible.
  - Revision of business models implemented.
  - The need for organizational restructuring.
  - Identification and completion of missing competencies.
  - Basic digital services to be transformed. This should also highlight the relationship between services (both mandatory and compulsory relations). Each service should also be assessed in terms of the level of digitalization possessed. Dependency of each module will point out the respective priorities.
  - Interface standards for data sharing and exchanges.
  - Filtering data to make a decision on certain aspects.
  - Data analysis and business intelligence in utilizing the data for better management.

Note that the proposed standards should be totally aligned with the standards published by standard authorizations if any.

- Performing assessments and commercialization of the prototypes: This should include: generating assessment criteria, receiving feedback from the users and generating improvement recommendations, making commercialization plans, receiving customer expectations and performing the technology watch in the research community as well as in the market.

- Enterprises may receive support from universities, research institutions and suppliers in order to generate required systems in accordance with the framework they developed. Note that the action plans generated in the light of this framework should also highlight the responsible person or groups as well as time to start and finish the activities.
- Identification of services to be digitized: During the transformation process nearly all functions of the enterprise need to be restructured with some possible digitalization. The enterprise may not have enough resources to run the process at a time for the overall enterprise as a whole. There might be prioritization among the services. Based on the strategic objectives and company priorities as well as the availability of resources, the services may be ordered. For each service, functions to be digitized can be listed. Technologic requirements and digital capabilities needed are to be listed. For each requirement is then turned into an action set for the assigned team and respective resource requirement plans are to be generated.
- While performing digitization, it is important to follow digitization standards if possible. If there are no standards available, the framework will provide the required guidelines.
- Forming the Implementation team: One of the main requirements in developing action plans is the involvement of senior executives and other leaders. Before starting with the action plan an implementation team should be formed. The main responsibility of the team is to transfer knowledge, provide support and enable existing staff and departments to adopt digital operations. Team members should be equipped with enough knowledge through training and consultancy to develop digital qualifications. Note that a implementation team for each project may be necessary in order to embed the proposed digitization in reality.

1.4.4 SHARING THE BEST PRACTICES

During the transformation it is extremely important to share the best practices. This may help increase motivation and provide encouragement for other projects to run. There are also some other benefits as listed below.

- Companies that share the best practices through knowledge sharing tools may adjust to any business change and improve employee performance.
- The best practices may help managers to recognize existing knowledge gaps and make improvement plans.
- Sharing the best practices may encourage the employees to share their innovative ideas.
- The best practices help ensure better and faster decision making.
- Sharing the best practices in an enterprise will considerably reduce the time spent looking for knowledge as all employees will have access to the right information by knowledge sharing.

- A validated set of knowledge can be stored and used whenever necessary.
- Using a knowledge sharing platform facilitates employees to learn business policies, practices and techniques.
- Best practices allow the know-how to be institutionalized.

1.5 CONCLUSION

Digital transformation is not a one-time phenomenon and it is a continuously running process. Nations, industries and enterprises should clearly understand the basic dynamics of this process and try to align themselves accordingly. This has been known and recognized by the academic and industrial communities for some time. An extensive literature review on Industry 4.0 and related technologies carried out by Oztemel and Gursev (2018) indicated, as a final remark, that it is now obvious that future manufacturing will be more intelligent, more flexible, more adaptive, more autonomous, more unmanned and more sensor-based. It is also revealed that more and more augmented reality will take place in manufacturing suites, naturally leading to the change of manpower profile as well. Another finding of this review is that the future manufacturing systems will not only be based upon Industry 4.0 standards as outlined today. They will be extended towards generating fully automated and unmanned systems with robots enriched with human-like behaviours. The correct use of real-time information is therefore expected to lead not only today but also the next industrial revolution.

Besides, it is now well known that the digital transformation will not only affect the manufacturing industries but all of society (education, trade, finance, food, health and law, etc.). To be able to cope with digital transformation requirements, it seems that companies should pay attention to certain interrelated social phenomena.

However, above all, there is no well-known standard and approach issued in order to carry out the transformation process. Each company should define its own strategy and roadmap concentrating on the available resources and urgency of the transformation needs. The need for research to make this process as easy as possible still remains. Universities and research centres should expend resources, time and effort to develop a set of standards to support as well as easy practical implementations.

REFERENCES

Basri, E. I., Abdul Razak, I. H., Ab-Samat, H., Kamaruddin, S. 2017. "Preventive Maintenance (PM) planning: A review". *Journal of Quality in Maintenance Engineering* 23(2). doi:10.1108/JQME-04-2016-0014.

Bhadani, A., Jothimani, D. 2016. "Big data: Challenges, opportunities and realities". In Singh, M. K., Kumar, D. G. (Eds.), *Effective Big Data Management and Opportunities for Implementation*, IGI Global, Hershey, PA, pp. 1–24.

Breining, C., Nunez, J., Alessi, M., Delahostria, E. G., Egenhofer, C., Haim, M., Hetzmannseder, E., Haryuki Ishio, H., Lanctot, P., Marchais, J. J., Mishima, H, Ogura, H., Paul, E., Rizos, V., Speh, R., Ueno, F., Xue, G., Zhang, D. 2014. *Orchestrating Infrastructure for Sustainable Smart Cities*, White Paper. Switzerland, International Electrotechnical Commission. Geneva, Switzerland.

Carlström, R., Aumund-Kopp, C., Riou, A., Coube, O., Murray, K. 2015. "Introduction to additive manufacturing technology". *Technical Report.* 1st Edition. European Powder Metallurgy Association, Shrewsbury, UK

Carmigniani, J., Furht, B. 2011. *Augmented Reality: An Overview, in Handbook of Augmented Reality* (edt by Furth B.) Springer Verlag, New York.

Furht, B. 2011. *Handbook of Augmented Reality.* ISBN: 978-1-4614-0063-9, New York, Springer Verlag. doi:10.1007/978-1-4614-0064-6.

Garnier, S., Jost, C., Jeanson, R., Gautrais, J., Asadpour, M., Caprari, G., Theraulaz, G. 2005. "Aggregation behaviour as a source of collective decision in a group of cockroach-like robots". In *Advances in Artificial Life*, LNAI 3630, eds. Mathieu S. Capcarrère, A. A. Freitas, P. J. Bentley, C. G. Johnson, J. Timmis, Springer Verlag, Berlin, Heidelberg, pp. 169–178.

Groß, R., Dorigo, M. 2008. "Evolution of solitary and group transport behaviors for autonomous robots capable of self-assembling". *Adaptive Behavior* 16(5): 285–305.

Grunow, O. 2016. *Smart Factory and Industry 4.0. The Current State of Application Technologies: Developing a Technology Roadmap*, Amazon Digital Services LLC, printed by Studylab, (e*book), ASIN: B077V56K19., ISBN 9783668271043,

Gursev, S. 2019. "Generating an assesment model for Industri 4.0". Unpublished PhD Dissertaiton, Institute of Pure and Applied Sciences. Marmara University

Hauert, S., Zufferey, J.-C., Floreano, D. 2008. "Evolved swarming without positioning information: An application in aerial communication relay". *Autonomous Robots* 26(1): 21–32.

Horváth, I., Gerritsen, B. H. M. 2012. "Cyber-physical systems: Concepts, technologies and implementation principles". In Horváth, I., Rusák, Z., Albers, A., Behrendt, M. (Eds.), *Proceedings of TMCE*, May 7–11, 2012. Vol 1, pp. 19–36.

Krishnaraj, N., Cherukuvada, S. 2018. "Virtual interaction". *International Journal of Information and Computing Science* 5(9): 335–355.

Maheswaran, M., Badidi, B. 2018. (Eds.), *Handbook of Smart Cities Software Services and Cyber Infrastructure*, Springer Verlag, Switzerland.

Nolfi, S., Floreano, D. 2000. *Evolutionary Robotics, Series on Intelligent Robots and Autonomous Agents.* MIT Press Cambridge, MA.

Oztemel, E. 2010. "Intelligent manufacturing systems". In Benyoucef, Lyes, Grabot, Bernard (Ed.), *Artificial Intelligence Techniques for Networked Manufacturing Enterprises Management.* Chapter 1. Springer Verlag. London.

Oztemel, E., Gursev, S. 2018. "Literature review of Industry 4.0 and related technologies". *Journal of Intelligent Manufacturing.*

Rafael, R. 2018. *Cloud Computing: From Beginning to End*, 2nd Edition, CreateSpace Independent Publishing Platform, Amazon, CA.

Rehman, A. A., Shrivastava, R. L. 2013. "Green manufacturing (GM): Past, present and future (a state of art review)". *World Review of Science, Technology and Sust. Development* 10(/2/3): 17–55.

Rose, K., Eldridge, S., Chapin, L. 2015. "The Internet of Things: An overview". *Technical Report*, Internet Society, Geneva, Switzerland, October 2015.

SeaLevel. 2019. *Industrial Automation Report*, April 2019. https://www.sealevel.com/2019/04/23/green-manufacturing-industry-4-0-and-the-sustainable-future/.

2 Digitalization in Industry
IoT and Industry 4.0

Gülçin Büyüközkan, Deniz Uztürk, and Öykü Ilıcak

CONTENTS

2.1 INTRODUCTION: BACKGROUND AND CURRENT SITUATION

Today we are moving towards an integrated manufacturing system, where every component is able to decide by itself as a single part while they can cooperate with other system participants (Gilchrist 2016). This creates a comprehensive system where every part is perpetually connected to the others. The connectivity contributes to the free data flow for every component, and this unstoppable data generate a valuable input for each and every participant to make some decisions, operate, and produce. The participants may be humans who play an essential role, or the system may be composed only by machines that can easily manage themselves. Regardless, the integrated system catalyzes to gather and analyze the significant amount of the data across machines. Hence, this enables faster, more flexible, and more efficient processes to produce higher quality goods with low costs (Oliver Scalabre 2019).

The idea of the fully connected system has emerged thanks to the fourth wave of technological advancement, which is called "Industry 4.0". This new buzzword has entered our lives rapidly thanks to its ability to create fast data flow and automation for our lives. However, mainly, Industry 4.0 has resulted due to the need for effective and waste-free mass production and manufacturing systems (Almada-Lobo 2016). Simply, Industry 4.0 could be defined as the adoption of automation and data exchange in manufacturing. Smart factories are the main goals for Industry 4.0, and it is a collective term that includes many contemporary automation systems, data exchanges, and production technologies. This revolution is a collection of values consisting of the internet of things (IoT), the internet of services (IoS), and cyber-physical systems (CPS).

However, this revolutionary system has not emerged out of the blue. It became possible as a result of the antecedent industrial developments. Every anterior revolution contributed to reaching today's automated systems. The primary history behind the digitalization in the industry can be summarized as follows (Klaus Schwab 2015):

The first industrial revolution (1.0) emerged with mechanical production systems using water and steam power. The second industrial revolution (2.0) introduced mass production with the help of electric power. In the third industrial revolution (3.0), the production was further automated with the digital revolution, the use of electronics, and the development of IT systems.

In detail, the innovation of steam machines enabled the construction of mechanical systems for manufacturing, and that speeded up the entire process. By the nineteenth century, developments led us to use the telephone and telegraph. Furthermore, Taylorism in the 1920s led the manufacturing systems to operate with more efficiency to produce more goods (Littler 1978).

In the system set by Taylor, all the movements of any work to be done were determined to the finest point, and the worker was asked to work as a machine. Briefly, the principles of Taylorism can be expressed as follows (Peaucelle 2000): First, all the necessary actions to determine the best conditions of the work are determined. Then they are told to the worker. Alongside this, the movements of the worker are strictly controlled during production activity. Taylorism, which emerged as a result of the mechanization movement, played a major role in the development of the industry. With this system, work efficiency increased. This approach concealed the possibility of mechanizing to obtain mass productions. Then, in the twentieth century, the developments of the computerized systems (the first micro-computer and Apple I) made it possible to automatize some part of the manufacturing systems. So, the computational processes laid the critical groundwork for further sophisticated automated systems. Figure 2.1 summarizes the development of the industrial revolutions over time.

Therefore, today, we are canalized to the automated manufacturing systems thanks to the IoT, digitalization, IoS, and cyber-physical systems (CPS). Connectivity stands at the centre of these primary systems.

CPS brings together virtual and physical worlds, enabling technologies that create a truly networked world in which intelligent objects communicate and interact with one another. When the boundaries between the real and virtual worlds disappear, systems that make and implement many innovative applications emerge (Lu 2017). The term is defined for Industry 4.0 itself since the new revolution represents the connection of the virtual and real worlds. Moreover, the IoT is the service that connects the devices, and owing to this connection, we give them nominal intelligence. However, IoS is in the crucial part of this information sharing. It visualizes and makes this bond valuable to create services (Wasmund 2017). Today, instead of copying the pioneering technology, manufacturers should think about their business model, and they need to ask how they can turn their products into services. For example, in the Netherlands, a catering company provides customized meals for each patient in hospitals considering the data generated from the hospital about their

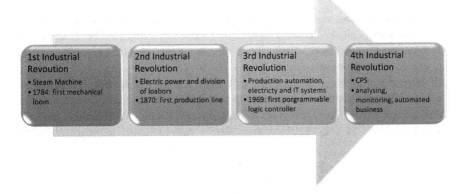

FIGURE 2.1 Development of the industrial revolution over time (Based on Zhou, Taigang Liu, and Lifeng Zhou 2015).

patient's needs (Wasmund 2017). So, IoS plays an essential role in customization of mass production, which can enable smart factories with flexible and smart products. In addition, the efficiency gained from this optimized data-based mass customization creates a revenue increase and profitability for the manufacturers.

The last but most crucial main principle of the industrial digitalization is the IoT, which is basically the connection of any device to the internet with an on/off switch. Accordingly, almost any product that anyone can think of can be included, from mobile phones, coffee machines, washing machines to headphones, lamps, wearable devices.

Nowadays, the internet is becoming more widespread, the cost of the connection decreases, Wi-Fi features and integrated sensors enable more devices, technology costs are reduced, and smartphones are in everyone's hands. All this creates a "perfect storm" for the IoT. Moreover, thanks to the IoT, new services are created to reach fully connected data-driven mass customization production systems.

So, the digitalization of the industry is driven by the existence of the internet. Production in the Industry 4.0 system is likened to a system in which machines serve and share information with products in real time. The German Research Center for Artificial Intelligence (DFKI) demonstrates how such a system will work in practice in a small intelligent factory in Kaiserslautern, Germany, which was founded with the contribution of 20 industrial and research partners, including Siemens. It uses soap bottles to demonstrate how products and manufacturing machines can communicate with each other. Empty soap bottles have radio frequency identification (RFID) labels on them, which enable machines to recognize the colour of the bottles. Thanks to this system, the information transmitted by radio signals of a product can be stored in digital media from the beginning of production. In this way, it emerges as a cyber-physical system (Glen White 2014).

Furthermore, Tesla is delivering vehicles with hardware and software that can be upgraded (Krasniqi and Hajrizi 2016). Their cars are sensor ready, and software upgrades enable the cars to have extra smartness delivered thanks to the internet.

Also, the customer can pay for the upgrades, which then generates bonus gains for Tesla.

On the other hand, another pioneer manufacturer, Otis, supplies elevators/lifts with sensors that can send data to their cloud platform (Claire Swedberg 2018). The data enables Otis to provide predictive maintenance service packages. Hence, they can add a durable revenue stream.

It is essential to build the strategy of a company on the deriving technologies since new developments bring various advantages such as efficiency and long-term revenue. Apart from these, Industry 4.0 and the IoT brings more benefits such as (Roblek, Meško, and Krapež 2016; Irniger 2018; Gilchrist 2016; Klaus Schwab 2015):

- Facilitates system monitoring and diagnostics.
- Self-awareness of systems and components.
- Environmentally friendly system and sustainable resource-saving behaviours.
- Higher productivity.
- Increases flexibility in production.
- Cost reduction.
- Development of new service and business models.

This chapter aims to provide a holistic review of the past and the current situation of Industry 4.0 under the frame of the IoT. For this purpose, the next section will provide the comprehensive benefits, opportunities, and the challenges for Industry 4.0. Afterwards, the sections will be more concentrated on the IoT to emphasize its importance in Industry 4.0. The main application area and case studies will be provided. Finally, the remarks will be given to conclude the subject.

2.2 INDUSTRY 4.0 AND THE IOT: STRENGTHS AND WEAKNESSES

The fourth industrial revolution, also known as "integrated industry", "intelligent manufacturing", is currently a trendy topic which has great potential to transform the traditional manufacturing systems to a new level. This need for smarter and flexible manufacturing systems is born out of the continuously growing international demand for capital and consumers products. This demand is also critical for human existence for its environmental, fiscal, and social dimensions (Stock and Seliger 2016). It obliges this consumption to exist as a human being in society, and this becomes a challenge for manufacturing systems. To overcome this challenge, the new industrial revolution provides some solutions by integrating all the systems, as mentioned in the previous section.

This development grants new opportunities through sustainability. Somehow, it has some constraints or weaknesses that can create risks for the manufacturers (Preuveneers and Ilie-Zudor 2017). As a first step, the beneficial areas can be generalized as (Zhou, Taigang Liu, and Lifeng Zhou 2015):

- Production transparency
- Information management

- Smart manufacturing
- Decision-making
- Optimization

Production transparency provides end-to-end control for the manufacturing system. Wireless sensors connected to the systems enable the ability to perpetually monitor the performance of the system (Li et al. 2017). This transparency can offer a robust and efficient orchestration of business processes. It enables us to gain real-time data about the plants, which can decrease the number of interruptions. However, this end-to-end real-time monitoring has some constraints. The main constraint is the infrastructural requirements. The network latency is the critical concern, and also it can result in high energy consumption (Zhou, Taigang Liu, and Lifeng Zhou 2015). Moreover, the wireless sensors can be challenging for the industrial environments where the humidity and temperature is variable compared to average values. Frequently changing environment may affect the reliability of the data collected from these devices.

These wireless technologies, primarily IoT-based technologies, are not only suitable for heavy industries or production of goods. They can also be useful in sectors such as healthcare, mine security, logistics, and firefighting (Xu, He, and Li 2014). However, there is a need for standardization of communication and identification systems. Also, the challenge for security and privacy is still on the agenda since the data is available to various systems users (Christin, Mogre, and Hollick 2010; Gaj, Jasperneite, and Felser 2013).

Optimization of the process is eased with the availability of Big Data and cloud computing since they enabled Machine-to-Machine (M2M) communication. However, the computational infrastructure in the site is on the whole not enough to handle these complex systems. The need for an extra resource (storage, software) has emerged to overcome the infrastructural challenges. For this purpose, cloud computing is one of the enabling technologies for Industry 4.0. The IoT allows the collection of data from various items; cloud computing allows storage and management of the data being collected from the devices. There is still a challenge while managing the high amount of data (O'Donovan et al. 2015).

Beside M2M communication, the humans are not yet taken out of the system completely. For strategic decisions, humans have to interact with virtual systems to solve problems. This cyber-physical environment helps strategic decision-makers to augment their ability of flexible analyzing and concluding (Gorecky et al. 2014).

In the light of these opportunities and challenges, the essential benefits and challenges are given in Table 2.1.

According to the applications, constraints, and the opportunities it is better to say that Industry 4.0 applications need a comprehensive approach that includes further complications for the development processes. The need for predictable system behaviour, quality assurance, and regulations are the crucial points to be taken care of while moving towards the fourth industrial revolution. Intelligent environments are not far in the future; they are applicable now with some challenges. New technological developments will help to diminish their risks to create more sustainable and efficient manufacturing and service systems thanks to the IoT's connectivity. The

TABLE 2.1

The Strengths and Weaknesses of Industry 4.0

Strengths	Weaknesses
Production efficiency	Complexity
Flexibility	Big amount of data
High revenue	Security
Real-time data	Privacy
Sophisticated customer services	
Perpetual monitoring	
Augmented human performance	

next section will present the main applications areas of the IoT in Industry 4.0 and their best-case applications from various sectors.

2.3   MAIN APPLICATIONS AND SECTORS OF THE IOT

The IoT has become a transformation technology for many areas by creating design innovation with new digital and intelligent technologies. The IoT has a wide range of applications from industrial automation, health, building and home automation to transportation and public services. The IoT provides applications that facilitate human life and operational processes in many areas and many different sectors. In this section, we examine the main applications and sectors of the IoT.

IoT applications are found in many different sectors in today's business world. Many companies fail to keep pace with the rapid transformation of technology. For this reason, it is vital to implement practices that will comply with the IoT in the sector (Elkhodr and Cheung 2016). Ten sectors which are frequently mentioned in the literature can be given as (Beecham Research 2016).

- Automotive: Automobiles connected to the IoT turn the data into actionable ideas both in and around the vehicle.
- Energy: With the IoT, countless networked devices can share information in real time, enabling more efficient distribution and better management of energy.
- Healthcare: The IoT reshapes the healthcare industry with solutions ranging from clinical wearable technologies to first-aid tablets and complex surgical equipment.
- Consumer & Home: IoT technology makes safe, smart homes a reality, from recognizing your voice to knowing the person at the door.
- Building: It provides solutions for increased energy costs, sustainability, and regulatory compliance by linking, managing, and securing devices that collect data from primary systems.
- Manufacturing: IoT technology enables factories to achieve operational efficiency, optimize production, and increase worker safety.

- Retail: The IoT offers unlimited opportunities for retailers to increase their supply chain efficiency, develop new services, and reshape the customer experience.
- Transportation: The IoT can save lives, reduce traffic, and minimize the environmental impact of vehicles, ranging from connected or driverless cars to intelligent transport and logistics systems.
- Logistics and Supply Chain: With the IoT, faster and more qualified deliveries can be achieved by keeping all elements of the supply chain in touch, and with real-time monitoring, smoother and safer processes can be achieved.
- Financial Services: The IoT can provide personalization and provide potential solutions that help customers better manage their financial conditions by collecting and analyzing data on customers' behaviour, spending patterns, and earnings.

Following the sectors that we have mentioned above, some examples of the areas where the IoT is found most frequently are given as follows.

The IoT's most essential application area can be given as smart cities. Smart cities, as the name implies, are the cities that integrate internet/technology in all these processes such as water, energy, traffic and waste management, education and health services. Improving the quality of life of people living in cities is the main objective of smart cities. IoT solutions in the smart city sector solve a variety of city-related problems that reduce traffic, air, and noise pollution and help make cities safer (Khan 2012).

Another prominent IoT application is smart home applications. A smart home is a house in which electrical devices are connected to a computerized control centre via a network, which can be remotely controlled or can automatically perform certain tasks under certain conditions. In other words, these are houses, which can ensure that the devices in the home can now operate at the right time without the need for human intervention. For example, illumination systems can be used to start or stop the lighting system of the house, as they see darkness and illumination through sensors. They adjust the louvre position according to the amount of light entering through the windows. Also, they start or stop operating the necessary light, music, heating systems in the rooms by using motion or sound detectors when a person enters or leaves the house (Arıkan, n.d.).

Another application area, wearable technologies, stands out as the smallest but functionally most useful and perhaps even the most comfortable member of the IoT world. Today, the most common examples are smart watches and fitness trackers (devices with functions such as daily running, step counter, calorie calculator). It has emerged as a sector that has been developing rapidly in recent years and increasing its product range (Team 2018).

Another essential application is smart grids. An intelligent grid promises to use the information obtained about the behaviour of consumers and suppliers to increase the efficiency, economy, and reliability of electricity distribution and provides monitoring and control of grids (Gour 2018).

The IoT has also gained important application areas in the health and medical sectors. Real-time or periodic health information (blood sugar, blood pressure, heart rate, body temperature, number of steps, instant physical status, etc.) of people with chronic illnesses or older people in need of care can be obtained from the relevant medical devices. The ability to monitor and analyze this information by the physician and to react immediately in severe situations can only be realized with IoT-based systems (Horn et al. 2016).

Another application area is agriculture. Agriculture is a fundamental necessity for human life and plays a significant role in the economy of countries. In order to cope with the problems in this area, it is crucial to ensure the integration of agriculture with technology. At this point, the IoT controls the environmental parameters in the agricultural sector, while increasing plant yield, and minimizing production cost and energy losses. Besides it provides more efficient processes. Plants can be monitored and controlled by embedded devices in greenhouses, while the climate in the greenhouse can be controlled. At the same time, the sensors can process data by measuring many different parameters according to the plant requirement (Gondchawar and Kawitkar 2016).

It can be said that one of the fastest developments for the IoT is the industrial area and the manufacturing sector. While providing a cost advantage with IoT applications, performance can be improved by increasing customer satisfaction. The purpose of the IoT in the industry can be defined as: real-time monitoring and control of processes, assigning smart machines, smart sensors, smart controllers with particular communication and internet technologies, high-precision automation and control to maximize safety and reliability (Ercan, Tuncay, and Kutay, Mahir 2016). In addition, the concept of Industrial IoT (IIoT) has been introduced in order to use the necessary data that can be transferred from industrial systems related to production/manufacturing in the information system. The information flow of IIoT causes changes that may positively affect our daily lives, business life, and industrial production systems. By combining industry and the IoT, the smart devices used in production minimize human error and enable real-time information to be evaluated by decision support systems, which will have many positive effects such as increasing production quality, reducing costs, and creating competitive products (Ercan, Tuncay, and Kutay, Mahir 2016).

Besides, the IoT provides innovative applications in transportation. Thus, control, data processing, and communication can be integrated between various transportation systems. Intelligent traffic control, autonomous navigation, vehicle communication, automatic transmission for emergency rescue, security and roadside assistance, intelligent parking systems, and many more examples can be given (Kosunalp and Arucu 2018) as application areas of the IoT in transportation.

As technology is continuously evolving at an increasing rate, there is no doubt that the IoT will continue to develop in the specified areas and many other areas. IoT applications will bring different perspectives to this world where we live and work, make our lives simpler and better than ever, and bring a revolution to the world. The future of the IoT will create a world where billions of things are interconnected and talk to each other, with minimal human intervention. In the next section, the future scenarios for the IoT will be examined.

2.4   FUTURE SCENARIOS OF THE IOT

With the rise of the IoT, the number and diversity of connected devices are expected to increase exponentially. IoT devices are becoming a part of the mainstream electronics culture, and people are adopting smart devices into their homes or businesses faster than ever. According to Gartner, the IoT will include 20 billion units installed by 2020 (Gartner and President 2017).

It is anticipated that IoT applications will further develop in the future, and the IoT will make significant progress with the increasing number of interconnected IoT devices around the world.

In the future, thanks to the IoT, operational technology and information technology will be merged, and more cost-effective connected sensors, artificial intelligence and control, faster and more widespread communication networks, cloud infrastructure and advanced data analysis capabilities will allow the fourth wave of the industrial revolution. In addition, previously unused/unprocessed data will be transformed into action with the IoT and become useful information that allows organizations to take their customer experience to the next level. In the future, the IoT will help countries respond to the biggest challenges that threaten our planet, such as global warming, water scarcity, and pollution, and will serve as a source of economic growth by continuing to provide innovative solutions for businesses, governments, and economies ("IOT 2020: Schneider Electric Issues Predictions Based on Global Study" 2016).

According to Statista statistics, 40 billion dollars will be spent on IoT systems by 2020 in each of the discrete production, transport and logistics, and public services areas. It is also foreseen that consumer electronics will constitute 63% of all IoT units installed in 2020 (Statista 2018). In other statistics of Statista, it is said that in 2025, there will be 75.44 billion connected devices of the IoT. The graph in years is given in Figure 2.2 (Privitera and Li 2018).

IoT applications will evolve over time and expand into more areas, but at the same time, it is anticipated that these applications will face and bring some challenges in the future. Of course, the most important of these is security problems caused by cyber-attacks. This risk can be seen, especially in low-security buildings and

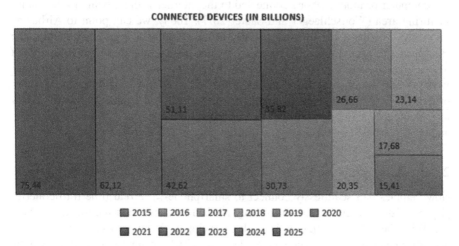

CONNECTED DEVICES (IN BILLIONS)

■ 2015 ■ 2016 ■ 2017 ■ 2018 ■ 2019 ■ 2020
■ 2021 ■ 2022 ■ 2023 ■ 2024 ■ 2025

FIGURE 2.2 Projected connected IoT devices worldwide from 2015 to 2025 (Based on Privitera and Li 2018).

organizations. IoT devices are vulnerable to attacks that could interfere with their regular operation, and this vulnerability is a fundamental problem that needs to be solved. However, most organizations do not want to pay more for tighter security. In this case, it is a risk that will prevent the growth of IoT applications (Deloitte 2018). Another challenge will arise from the need for infrastructure to connect billions of IoT devices in the future. Existing central systems will need massive investments to overcome large amounts of information. Another challenge that the IoT should take into consideration in the future is the lack of standards. Lack of standards can slow down the adoption of the IoT and prevent the development of new technologies (PwC 2016). While taking new steps for the IoT in the future, these difficulties and risks should be taken into consideration, and the right decisions should be made in this direction.

2.5   CASE STUDIES OF THE IOT

As we mentioned in previous sections, the IoT leads to change in many sectors such as logistics, health, manufacturing, retail, energy, and agriculture. Many companies now use the IoT to gain a competitive advantage and improve their processes and to pursue a more innovative approach. This section gives real-life examples of the use of IoT technologies from main three industries.

2.5.1   MANUFACTURING INDUSTRY

In general, the IoT in the manufacturing area is the application of instrumentation, connected sensors, and other devices to machines and processes. The communication of the smart devices used in production or industrial areas and the control of the data resulting from this communication from a single device include the activity of IoT applications in the manufacturing area. This control can be accessed from the computer or mobile phone connected to the internet without being in the manufacturing area (Wollschlaeger and Sauter 2017). Today, we can point to Airbus as an example of using the IoT successfully in the manufacturing sector. Airbus, the European aircraft manufacturer, not only applies IoT technologies in its factories but also applies these tools to the tools used by its employees in manufacturing. In other words, while establishing the factories of the future, it also constitutes the workers of the future. Workers in the factory of the future can use a tablet or smart glasses to evaluate a task and then transfer this information to the robots. Thus, much faster and healthier business processes are created (Saran 2019).

2.5.2   AUTOMOTIVE INDUSTRY

Now vehicles can seamlessly connect to smartphones, see real-time traffic alerts, play playlists, and connect to emergency roadside assistance at the touch of a button. With the IoT, this situation is further developed. Thanks to sensors, the internet, advanced software, and connected devices, vehicles can now communicate with each other or with the user to exchange data (Durukan, n.d.). Cars with sensors that provide real-time information flow can measure gas/oil/liquid levels and engine temperature, enable rear-view cameras to steer and direct proximity sensors for parking. For example, as a real-life application, connected tools can process and share real-time data, enabling faster service delivery (Deloitte University Press 2015).

An example of this is when the car sends a calendar appointment to the nearest service. Today, accidents can be minimized with the help of the cars' sensors. Tesla, one of the giants in the automotive sector, is developing IoT-based autonomous driving systems in conjunction with cloud technology to build driverless automobiles today. Also, BMW attaches importance to connected vehicles to integrate vehicle-related services (Deloitte 2017).

2.5.3 HEALTHCARE INDUSTRY

As mentioned in the previous sections, the IoT can be used in healthcare in many ways, from the collection of health data of patients to various mobile medical applications. An example of this in real life is the "elderly care practice". With this application, IoT devices allow continuous monitoring of the health data of elderly patients, while ensuring that the location of older people with diseases such as Alzheimer's, who are therefore at risk of disappearance, is known at any time (Altan 2017). Besides, another application focuses on collecting and analyzing the health data. Collection of health data using IoT technology can make treatment processes much easier and more efficient for both patients and doctors. In this way, health data can be collected, and the course of the disease can be monitored wherever patients are. This can offer a great benefit, especially in the treatment of people with chronic illnesses, and provides rapid response to the disease if their disease progresses adversely. In this way, the chances of patient survival for fatal diseases are significantly increased. At the same time, with the use of the IoT in the field of health, the necessity of keeping the patients in the hospital is eliminated. Thanks to various health devices connected to the internet, the status of patients with serious illnesses can be monitored remotely by doctors and nurses (Econsultancy 2019).

Apart from the industries mentioned, the IoT is developing in many different industrial areas and creating smarter and innovative processes. By using this technology business processes are provided to be more error-free, faster, with higher customer satisfaction, and more cost-effective. In this study, the definition of historical development of the IoT in the context of Industry 4.0 and its main application areas/sectors are mentioned. The development of the IoT will bring more convenience in the future, as well as several challenges. In this context, the future of the IoT is discussed in this study and the future scenarios of the IoT are mentioned by making use of the studies put forward in the research reports on this subject. Afterwards, the study was completed by giving real-life examples. This chapter aims to provide necessary information about the IoT to both researchers and managers and to emphasize the place and importance of the subject at present.

2.6 CONCLUDING REMARKS

In this age, where we have to deal with large-scale masses of information generated from multiple channels, the rules of the game are rewritten and those who can rebuild business processes can successfully integrate with the digital ecosystem for production. In short, those who are able to adapt to the changes and transformations in digital technologies will be able to remain both corporate and individual. Organizations need to embrace Industry 4.0 in order to meet the needs of their customers and provide a competitive advantage over their competitors. Market dynamics will change, and those who can meet customer needs quickly and accurately will survive.

Industry 4.0 brings new career opportunities with economic growth. One of the most critical requirements of this era is the human resources with the necessary skills. However, this is the most critical feature expected from human resources: to be able to have interdisciplinary skills rather than specialized skills with analytical competencies. It is vital that individuals acquire the necessary skills to cope with the digital work environment.

Industry 4.0 brings some challenges with numerous advantages. Above all, being able to adapt to the technologies within the scope of Industry 4.0 and enable the institutions to turn them into a corporate culture, and most importantly, to provide the necessary human resources within the organization are the primary adaptation steps for the new developments. We can say that we are still in the early days of Industry 4.0. There are challenges on many fronts, such as the integration of virtual and physical worlds, data capabilities, implementation challenges, guidelines and strategic capacities, skills, culture, standards, and levels of maturity/readiness in these areas.

It is imperative to be ready for the new digital capabilities emerging with Industry 4.0. We have to create an effective strategy for ourselves and find a way in the Industry 4.0 era by linking it to our personal goals. With Industry 4.0, markets will continuously change, and expectations will become more fluid. While all this is happening, individuals and organizations will need to think and act on their survival strategies.

Manufacturers carry out a wide range of activities in research and development (or engineering), IT, manufacturing, logistics, marketing, sales, services, human resources, procurement, and finance. New features of smart connected products change every activity in this value chain. What is reshaping the value chain is at the core of the data. People in smart factories still play a critical role in the manufacturing process. As increased operators, they control and monitor production sequences in the production network. There is no need for employees to be physically present at the factory. With IT-based help systems such as data glasses, an operator can manage a real factory (via extended reality) almost remotely. These assistance systems can also be tailored to the individual abilities and needs of staff members and have the potential to extend the retention of the elderly in the workforce.

In this chapter, we aimed to address the past and present situation of the new industrial revolution by highlighting the IoT, which is the foundation of Industry 4.0 in particular. We talked about the advantages it has and the limitations it faces. In the last sections, we present the main application areas and cases of the IoT, which are the basis of Industry 4.0.

As a result, the new industrial revolution, even if it involves some risks, has excellent benefits in terms of productivity and sustainability and is likely to do so in the future. With the integrated approach to Industry 4.0, the digitalization of sectors will be stronger and more sustainable. New technologies will also pave the way for mitigating the risks today.

REFERENCES

Almada-Lobo, Francisco. 2016. "The Industry 4.0 Revolution and the Future of Manufacturing Execution Systems (MES)." *Journal of Innovation Management* 3(4): 16–21. doi:10.24840/2183-0606_003.004_0003.

Altan, Serdar. 2017. "Sağlık Alanındaki Nesnelerin İnterneti Uygulamaları." https://blog.iven.io/sa%C4%9F1%C4%B1k-alan%C4%B1ndaki-nesnelerin-i%CC%87nterneti-uygulamalar%C4%B1-760842532f9e.

Arıkan, Işıl. n.d. "Nesnelerin İnterneti (IoT) VII: Giyilebilir Teknolojiler." www.fiyatimbu.com/blog/Single/320.

Beecham Research. 2016. "World of Connected Services Internet of Things." http://www.beechamresearch.com/article.aspx?id=4.

Christin, Delphine, Parag S. Mogre, and Matthias Hollick. 2010. "Survey on Wireless Sensor Network Technologies for Industrial Automation: The Security and Quality of Service Perspectives." *Future Internet* 2(2): 96–125. doi:10.3390/fi2020096.

Claire Swedberg. 2018. "IoT Elevates Service and Lowers Downtime for Otis Customers - 2018-07-11 - Page 1 - RFID Journal." https://www.rfidjournal.com/articles/view?17648.

Deloitte. 2017. "The Internet of Things: Revolution in the Making." https://www.nasscom.in/natc2017/images/white-papers/internet-of-things.pdf.

Deloitte. 2018. "IoT Innovation Report." https://www2.deloitte.com/content/dam/Deloitte/de/Documents/Innovation/Internet-of-Things-Innovation-Report-2018-Deloitte.pdf.

Deloitte University Press. 2015. "Who Owns the Road? The IoT-Connected Car of Today—and Tomorrow." https://www2.deloitte.com/content/dam/Deloitte/fr/Documents/strategy/dup_IoT-who-owns-the-road.pdf.

Durukan, Işıl. n.d. "Nesnelerin İnterneti (IoT) VI: Bağlantılı Arabalar." https://www.fiyatimbu.com/blog/nesnelerin-interneti-iot-vi-baglantili-arabalar_319.

Econsultancy. 2019. "10 Examples of the Internet of Things in Healthcare." https://econsultancy.com/internet-of-things-healthcare/.

Elkhodr, Mahmoud, and Hon Shing Cheung. 2016. "Internet of Things Applications: Current and Future Development." In *Innovative Research and Applications in Next-Generation High Performance Computing*, 397–427. doi:10.4018/978-1-5225-0287-6.ch016.

Ercan, Tuncay, and Mahir Kutay. 2016. "Internet of Things (IoT) Applications in Industry." *Afyon Kocatepe University Journal of Sciences and Engineering* 16(3): 599–607.

Gaj, P., J. Jasperneite, and M. Felser. 2013. "Computer Communication Within Industrial Distributed Environment—a Survey." *IEEE Transactions on Industrial Informatics* 9(1): 182–189. doi:10.1109/TII.2012.2209668.

Gilchrist, Alasdair. 2016. *Industry 4.0: The Industrial Internet of Things*. Berkeley, CA: Apress.

Glen White. 2014. "INDUSTRY 4.0: German Manufacturers Introduce Smart Factories to Spark Growth." *Global Manufacturing*, October 30. https://www.manufacturingglobal.com/technology/industry-40-german-manufacturers-introduce-smart-factories-spark-growth.

Gondchawar, Nikesh, and Prof R.S. Kawitkar. 2016. "IoT Based Smart Agriculture." *International Journal of Advanced Research in Computer and Communication Engineering* 5(6): 838–842. doi:10.17148/IJARCCE.2016.56188.

Gorecky, D., M. Schmitt, M. Loskyll, and D. Zühlke. 2014. "Human-Machine-Interaction in the Industry 4.0 Era." In *2014 12th IEEE International Conference on Industrial Informatics (INDIN)*, 289–294. doi:10.1109/INDIN.2014.6945523.

Gour, Rinu. 2018. "Top 10 Applications of IoT." https://dzone.com/articles/top-10-uses-of-the-internet-of-things.

Horn, Geir, Frank Eliassen, Amir Taherkordi, Salvatore Venticinque, Beniamino Di Martino, Monika Bücher, and Lisa Wood. 2016. "An Architecture for Using Commodity Devices and Smart Phones in Health Systems." IEEE *Workshop on ICT Solutions for EHealth*. doi:10.1109/ISCC.2016.7543749.

Hung, M. 2017. "Gartner Insights on How to Lead in a Connected World." https://www.gar tner.com/imagesrv/books/iot/iotEbook_digital.pdf

"IOT 2020: Schneider Electric Issues Predictions Based on Global Study." 2016. https://www.globalbankingandfinance.com/iot-2020-schneider-electric-issues-predictions-based-on-global-study/.

Irniger, Anna. 2018. "From IoT to IIoT to Industry 4.0: The Evolution of the Industrial Sector." Accessed October 28. https://www.coresystems.net/blog/from-iot-to-iiot-the-evolution-of-the-industrial-sector.

Khan, Rafiullah, Sarmad Ullah Khan, Rifaqat Zaheer, and Shahid Khan. 2012. "Future Internet: The Internet of Things Architecture, Possible Applications and Key Challenges." In *10th International Conference on Frontiers of Information Technology (FIT): Proceedings*, 257–260, Islamabad, India.

Kosunalp, Selahattin, and Muhammet Arucu. 2018. "Nesnelerin Interneti ve Akıllı Ulaşım." *Akıllı Ulaşım Sistemleri ve Uygulamaları Dergisi* 1(1): 1–7.

Krasniqi, X., and E. Hajrizi. 2016. "Use of IoT Technology to Drive the Automotive Industry from Connected to Full Autonomous Vehicles." *IFAC-PapersOnLine*, 17th IFAC Conference on International Stability, Technology and Culture TECIS 2016, 49(29): 269–274. doi:10.1016/j.ifacol.2016.11.078.

Li, Xiaomin, Di Li, Jiafu Wan, Athanasios V. Vasilakos, Chin-Feng Lai, and Shiyong Wang. 2017. "A Review of Industrial Wireless Networks in the Context of Industry 4.0." *Wireless Networks* 23(1): 23–41. doi:10.1007/s11276-015-1133-7.

Littler, Craig R. 1978. "Understanding Taylorism." *The British Journal of Sociology* 29(2): 185–202. doi:10.2307/589888.

Lu, Yang. 2017. "Industry 4.0: A Survey on Technologies, Applications and Open Research Issues." *Journal of Industrial Information Integration* 6(June): 1–10. doi:10.1016/j.jii.2017.04.005.

O'Donovan, Peter, Kevin Leahy, Ken Bruton, and Dominic T. J. O'Sullivan. 2015. "Big Data in Manufacturing: A Systematic Mapping Study." *Journal of Big Data* 2(1): 20. doi:10.1186/s40537-015-0028-x.

Peaucelle, Jean-Louis. 2000. "From Taylorism to Post-Taylorism: Simultaneously Pursuing Several Management Objectives." *Journal of Organizational Change Management* 13(5): 452–467. doi:10.1108/09534810010377426.

Preuveneers, Davy, and Elisabeth Ilie-Zudor. 2017. "The Intelligent Industry of the Future: A Survey on Emerging Trends, Research Challenges and Opportunities in Industry 4.0." *Journal of Ambient Intelligence and Smart Environments* 9(3): 287–298. doi:10.3233/AIS-170432.

Privitera, Donald, and Lei Li. 2018. "Can IoT Devices Be Trusted? An Exploratory Study." 5. Paper 44. New Orleans, LA. https://aisel.aisnet.org/amcis2018/Security/Presentations/44/

PwC. 2016. "The Industrial Internet of Things." https://www.pwc.com/gx/en/technology/pdf/industrial-internet-of-things.pdf.

Roblek, Vasja, Maja Meško, and Alojz Krapež. 2016. "A Complex View of Industry 4.0." *SAGE Open* 6(2): 2158244016653987. doi:10.1177/2158244016653987.

Saran, Cliff. 2019. "How Airbus Handles IoT Network Traffic in Manufacturing." https://www.computerweekly.com/news/252456522/How-Airbus-handles-IoT-network-traffic-in-manufacturing.

Scalabre, Oliver. 2019. "Industry 4.0 - the Nine Technologies Transforming Industrial Production." Boston Consulting Group. https://www.bcg.com/en-us/capabilities/operations/embracing-industry-4.0-rediscovering-growth.aspx.

Schwab, Klaus. 2015. "The Fourth Industrial Revolution What It Means and How to Respond." *Foreign Affairs*, December 12. http://www.inovasyon.org/pdf/WorldEconomicForum_ The.Fourth.Industrial.Rev.2016.pdf.

Statista. 2018. "Internet of Things Spending Worldwide by Vertical 2015 and 2020." https:// www.statista.com/statistics/666864/iot-spending-by-vertical-worldwide/.

Stock, T., and G. Seliger. 2016. "Opportunities of Sustainable Manufacturing in Industry 4.0." *Procedia CIRP*, 13th Global Conference on Sustainable Manufacturing – Decoupling Growth from Resource Use, 40(January): 536–541. doi:10.1016/j.procir.2016.01.129.

Team, Dataflair. 2018. "IoT Applications | Top 10 Uses of Internet of Things." https://data-flair.training/blogs/iot-applications/.

Wasmund, Ryan. 2017. "The Internet of Services in Industrie 4.0." *Concept Systems Inc.* May 16. https://conceptsystemsinc.com/the-internet-of-services-in-industrie-4-0/.

Wollschlaeger, Martin, and Thilo Sauter, and Juergen Jasperneite. 2017. "The Future of Industrial Communication." IEEE Industrial Electronics Magazine 11 (1): 17–27. doi:10.1109/MIE.2017.2649104.

Xu, L. D., W. He, and S. Li. 2014. "Internet of Things in Industries: A Survey." *IEEE Transactions on Industrial Informatics* 10(4): 2233–2243. doi:10.1109/TII.2014.2300753.

Zhou, K., Taigang Liu, and Lifeng Zhou. 2015. "Industry 4.0: Towards Future Industrial Opportunities and Challenges." In *2015 12th International Conference on Fuzzy Systems and Knowledge Discovery (FSKD)*, 2147–2152. doi:10.1109/FSKD.2015.7382284.

3 Autonomous Robots and CoBots

Applications in Manufacturing

Miguel Ángel Moreno, Diego Carou, and J. Paulo Davim

CONTENTS

3.1 DESCRIPTION OF THE TECHNOLOGY IN THE CONTEXT OF INDUSTRY 4.0

Industry 4.0 (I4.0), also known as the Fourth Industrial Revolution, is a large German strategic programme of the federal government with universities and private companies (Kagermann, 2013; Xu, Xu and Li, 2018). A similar approach to this new manufacturing philosophy was promoted by the United States with the Smart Manufacturing initiatives (SMLC, 2011). They both resort to smart approaches to working activities in the value chain (Longo, Nicoletti and Padovano, 2017; Stock *et al.*, 2018). If previous Industrial Revolutions were due to steam power, electricity and information, this Fourth Revolution is characteristic for Cyber-Physical Systems (CPS).

A CPS (Lee, 2006; Baheti and Gill, 2011) is an essential concept in I4.0 referring to entities where "the physical and virtual world grow together" (Thoben, Wiesner and Wuest, 2017). It refers to a system that collects data of itself and environment, processes and evaluates these raw data, exchanges information with other systems, makes local decisions and initiates actions by itself. To this end, it counts on sensors, wireless communication, computer processing and actuators (Klocke *et al.*, 2011; Wang, Törngren and Onori, 2015). In addition to the previous features, there are some other interesting aspects of CPS (Cengarle *et al.*, 2013): control algorithms for self-organization, adaptability to changing conditions, new efficient communications across cross domains, embedded and IT dominated systems and human outside or inside the loop. For the latter, autonomous robots and collaborative robots (CoBots) are examples of the absence and presence of the human in the loop, respectively.

Even though everyone apparently knows what a robot is, it is convenient to give an accurate definition (Ben-Ari and Mondada, 2017). According to the Oxford English Dictionary, a robot is "a machine capable of carrying out a complex series of actions automatically, especially one programmable by a computer". A crucial feature of robots which is not stated in the previous definition is the adoption of sensors, unlike automata which cannot adopt their actions to their environments.

Robots in manufacturing are a type of CPS essential to get flexibility, to shift from mass to customized production, and to adapt to technical evolution (Pedersen *et al.*, 2016; Alcácer and Cruz-Machado, 2019). Certain approaches state that robots can be considered as one of the forms of Artificial Intelligence (AI) (Wu, Liu and Wu, 2018). Considering them as a reconfigurable automation technology with AI, robots are an attractive technology as their adaptability can facilitate different product manufacturing and, therefore, decrease product costs (Salkin *et al.*, 2017). In other words, robots customize products at large scale, with better resource consumption and under changing conditions (Wang *et al.*, 2016; Schuh *et al.*, 2017; Dalenogare *et al.*, 2018; de Sousa Jabbour *et al.*, 2018).

In 1942, Isaac Asimov designed three laws of robotics. Despite originally being proposed as a science-fiction matter, they are interesting reference guidelines to minimize potential risks and uncertainty in the I4.0 (Magruk, 2016).

Regarding the role the human worker plays, robots can be classified into autonomous robots and CoBots. On the one hand, autonomous robots are characterized by the following features (Alcácer and Cruz-Machado, 2019): computing, communication, control, autonomy and sociality, which are accomplished when microprocessors and AI are combined with goods, services and equipment to make them cleverer. Indeed, fully autonomous robots make their own judgements to execute tasks on dynamically variable atmospheres without operators' interaction (Ben-Ari and Mondada, 2017). For this reason, dirty and hazardous industrial applications on anarchic surroundings can be enhanced by an Autonomous Industrial Robot (AIR) or several ones in association. On the other hand, CoBots, unlike autonomous robots, introduce proximity of robots with humans in order to work together in flexibility-required tasks (Koch *et al.*, 2017). As some authors state: "For a smart robotics factory within the context of I4.0 and IoT, [...] collaboration between human workers and robots is the key" (Thoben, Wiesner and Wuest, 2017). In other words, it is usually referred to as a symbiotic relationship fed by human flexibility and machine accuracy. This connection keeps humans in the loop of data and information (Wang, Törngren and Onori, 2015). Cooperation between human–robot is possible because: (i) safety is guaranteed by limited on speed and force movements of the robot, which helps guide it by hand (Weiss and Huber, 2016); and (ii) the human–robot barrier is demolished, hence they are more affordable and flexible (El Makrini *et al.*, 2018). Active collision avoidance, dynamic task planning and adaptive robot control are endemic features of this type of robot (Wang, Törngren and Onori, 2015). In addition, great efforts are being developed to create common standards (e.g. ISO 10218) to guarantee safety and security for humans, facilities and equipment by the International Organization for Standardisation (ISO) (Brending *et al.*, 2017).

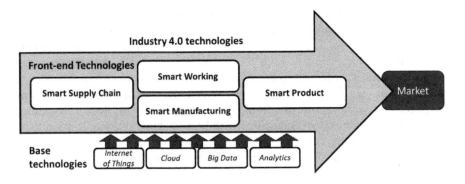

FIGURE 3.1 Theoretical structure of I4.0 technologies. Reprinted with permission from Elsevier (Frank, Dalenogare and Ayala, 2019).

According to Frank, Dalegonare and Ayala (2019), I4.0 technologies are divided into Front-end Technologies (Smart Supply Chain, Smart Working, Smart Manufacturing and Smart Product) and Base Technologies (Internet of Things, Cloud, Big Data and Analytics) (Figure 3.1). Whereas Front-end Technologies refer to final goals of I4.0, Base Technologies offer connectivity and brainpower to the Front-end Technologies. In other words, the latter are the driving forces connecting the real environment to the digital one. By means of these information and communication technologies (Elisabetta, 2016), a loop with data gathered in real time is created, which provides useful information for the manufacturing system making manufacturing lines more flexible (i.e. the use of sensors for predictive maintenance (Tao *et al.*, 2018)).

Internet connection with objects without human interaction, commonly known as the Internet of Things (IoT) (Nord, Koohang and Paliszkiewicz, 2019), is the foundation of the intercommunication of robots. In this way, every single one can be tracked and its dynamic information and data can be acquired at any time, as well as further applications such as controlling.

Cloud manufacturing, defined by Wang (2014) as "an integrated cyber-physical system that can provide manufacturing services digitally and physically to best utilize manufacturing resources" closes the loop from sensing, monitoring, planning and control. Four are the main cloud manufacturing services oriented to robotics (Wang, 2014): (i) functionalities for remote monitoring and control, (ii) distributed process planning, (iii) model-guided remote assembly and (iv) active collision avoidance for human–robot collaboration. An interesting instance of data in the cloud is the integration of the lifecycle of the product, as well as its supply chain activities, to the manufacturing process (Wang *et al.*, 2016; Dalenogare *et al.*, 2018). All in all, it could be said that items deliver their own production data to intelligent robots (Shirase and Nakamoto, 2013). Another objective in the context of cloud manufacturing addresses the issue of minimizing energy consumption of robots during assembly in a cloud environment (Wang *et al.*, 2018): given a robot path and based on its kinematics and dynamics models on the cloud, the path and configuration for the minimum energy consumption is determined and chosen.

The continuous growth of Big Data (BD) in industrial IoT is evident due to the massive deployment of sensors and IoT devices, in such a way that Big Data Analytics (BDA) is the foundation of operational intelligence of robots (Rehman *et al.*, 2019). Manufacturing Execution Systems (MES), which are computer-based systems utilized in production to keep a record of the conversion of raw supplies to finished goods, help keep the track record of all manufacturing data in a live way and gather up-to-date information from robots (Ariyaluran *et al.*, 2018; Mohammed *et al.*, 2018). In addition, MES (together with BDA) help production decision makers comprehend how present circumstances and processes on the plant floor can be adjusted to enhance manufacturing output (Rehman *et al.*, 2019).

3.2   HISTORICAL EVOLUTION: ORIGINS AND CURRENT STATE. STRENGTHS AND WEAKNESSES

Origins of self-organizing manufacturing, which includes the notions of agility, adaptability, efficiency, reorganization and self-organization (J. Zhang *et al.*, 2017), dates back to the beginning of the 21st century, when initial work towards new manufacturing and assembly systems proved the idea of industrial agents (Rizzi, Gowdy and Hollis, 2001). The term agent refers to a "mechanically, computationally, and algorithmically modular manufacturing entity (e.g. robot) capable of both communication and physical interaction with its peers". Every single agent would be part of a larger architecture for cooperative behaviours.

Rochwell Automation (Maturana and Norrie, 1996) and Schneider Electronics (Colombo, Schoop and Neubert, 2006) developed interesting solutions with a generic multi-agent mediator and an agent-based smart control platform, respectively. Further work was performed by academia (Monostori, Váncza and Kumara, 2006; Ribeiro, Barata and Ferreira, 2010). In 2011, the first self-organizing robotic assembly system grounded on Agent-Oriented Architectures (industrial agents on the core of the system) was proven at FESTO, Germany. This swarm configuration brought computer science to an industrial level, constituting smart sets of communicated industrial agents (Figure 3.2). Further I4.0 breakthroughs are continuously introduced at industrial fairs such as the Hannover Messe, which takes place every year.

On the one hand, current autonomous and CoBots within I4.0 are undoubtedly advantageous for the following reasons. First of all, robots enable manufacturing systems to reach mass customization, growth of productivity, adaptability and even rapidity of production, as well as a decrease in inventory (Gerbert *et al.*, 2015). Second, they allow the breaking down of traditional quality controlling through samples. Instead, an error-connecting machinery that adjusts production is implemented, hence a reduction in costs and the resulting increase in competitiveness (Bahrin *et al.*, 2016). Some studies claim that the top 100 European producers would save the expenses of scrapping and reprocessing if they could eradicate all flaws (Gerbert *et al.*, 2015). Third, a better integration of robots into human workspace becomes more cost-effective and efficient. I4.0 technologies capacitate robots to be controlled remotely, in such a way that they would text message a warning if any issue came up. Therefore, a plant could operate 24 hours a day while employees

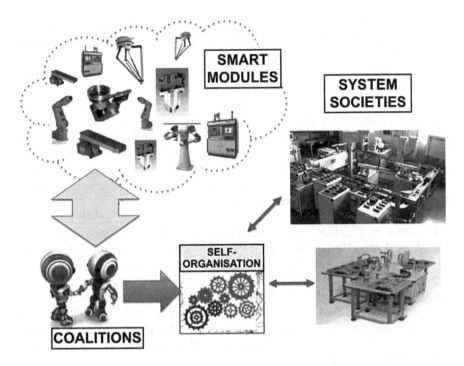

FIGURE 3.2 Modular blocks generation "system societies". Reprinted with permission from Elsevier (Wang, Törngren and Onori, 2015).

operate during the daytime only (Blanchet *et al.*, 2014). Another advantage related to human integration is that workers using CoBots are less prone to fatigue.

In terms of security, I4.0 plays a major role regarding safety in human–robot interaction (Thoben, Wiesner and Wuest, 2017). This integration is performed by means of stereo vision and advanced collision avoidance systems to build a safer and smarter working area (Liu *et al.*, 2000). Since robots and CoBots are able to sense, interpret and react, they dynamically activate protective mechanisms before a hazard; in such a way that a smart and dynamic security area focused on the worker is created. To that end, algorithms for image recognition and human movement prediction are to be implemented (Surdilovic, Schreck and Schmidt, 2010; Liu and Wang, 2018). Other methods in this direction include a vision-based approach such as 3D surveillance (Krüger *et al.*, 2005) via motion, colour and texture analysis, inertial sensor-based approach (Corrales, Candelas and Torres, 2011) via a motion capture suit, Time-of-Flight (ToF) method (Schiavi, Bicchi and Flacco, 2009) and depth camera-based approach (Fischer and Henrich, 2009; Flacco *et al.*, 2012).

On the other hand, I4.0 robots are subjected to several issues that hinder their establishment. A useful classification in technological and methodical weaknesses is followed (Thoben, Wiesner and Wuest, 2017).

With respect to technological issues, a major problem for the establishment of robots in the scope of I4.0 arises from the current heterogeneity of interfaces, standards and sensors of robots, hence the difficulty in entities' connection and interoperability.

To solve this, respected standards should be implemented. Nevertheless, this finds resistance in the fact that each manufacturing industry keeps different standards according to its interests.

Another weakness is the increase in the amount of raw data (from the sensors installed on the robots) and handling it by means of BDA (Y. Zhang *et al.*, 2017; Rehman *et al.*, 2019). This increases the complexity of the architecture, which affects data quality, validation, verification and communication routines. In fact, adjusting BD algorithms to manufacturing realities becomes a more difficult challenge (i.e. the need for automated data quality monitoring algorithms). As a consequence of the IoT, cloud computing and all the other I4.0 technologies, the data security issue comes up as an important concern where manufacturing data is prone to be stolen, and equipment is exposed to cyber-sabotage.

Related to methodological issues, the lack of reference models calls for coordination efforts to establish common definitions of fundamental concepts and modelling formalisms (Davis *et al.*, 2015; Papazoglou, van den Heuvel and Mascolo, 2015; Ho and O'Sullivan, 2017). In search of a solution, some authors propose ways of dynamic synchronization of decisions and action workflows in heterogeneous environments (Davis *et al.*, 2015; DIN/DKE). As the manufacturing industry is so diverse and each stakeholder has their own requirements on the user interface and information content, visualization comes up as a further challenge (Brodsky *et al.*, 2016). Therefore, it is necessary to break down all the raw data gathered by robots and present them in a suitable way.

3.3   MAIN APPLICATIONS AND SECTORS

Autonomous robots can be classified regarding their mobility in two categories: fixed and mobile robots (Ben-Ari and Mondada, 2017). Fixed robots are mainly industrial robotic operators that act in accurately defined surroundings tailored for robots, which simplifies its design. Industrial robots execute specific recurring tasks (i.e. welding or spraying parts in automotive manufacturing plants, a pioneer sector in robotic application). By contrast, mobile robots are designed in order to shift around and run chores in big, barely described and unclear environments which are not specifically intended for robots. They have to cope with circumstances which are not accurately known in advance and that vary over time.

Further approaches classify robots according to the environment in which they operate and according to the intended application field (see Figure 3.3) (Alcácer and Cruz-Machado, 2019). Whereas the first classification (aquatic, airborne, terrestrial and amphibious) is based on the operating environment, the second one focuses on the actual application for which the robot is designed (industry, services, raw material obtaining, etc.).

As identified in Figure 3.3, robotic applications exceed manufacturing applications by far. So, other applications should be analyzed because some of them could exert valuable improvements for manufacturing ones. For instance, an important application of robotic applications in a non-industrial field that is expected to gain worldwide attention in the coming years is the medical sector. Robotic-arm assisted technology is expected to be used in a wide number of interventions.

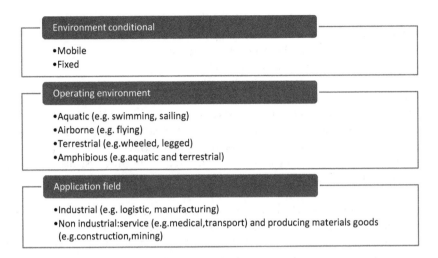

FIGURE 3.3 Characterization scheme for autonomous robots (Based on Alcácer and Cruz-Machado, 2019).

For instance, as stated by Deese *et al.* (2018), the use of robotic-arm assisted technology for unicondylar knee arthroplasty (UKA) has even revolutionized the procedure based on results of interventions carried out by using Stryker's Mako™ robotic arm. Attention should be also given to state-of-the-art solutions by advanced organizations such as NASA. They are developing humanoid robots such as Robonaut 2 and Valkyrie in collaboration with General Motors and Oceaneering that are required to interact with astronauts in space environments (Balint and Hall, 2015).

Regarding manufacturing industrial robots, the assembly line, ranked third after handling and welding robots, "is more productive than ever thanks to robotics" (Nye, 2013). In 2018, 75560 industrial robots were sold in Europe, where Germany is by far the largest market with 26723, mainly due to installations by car producers. However, Asia and Australia are the leaders in terms of installations of industrial robots numbering 283000, with China being the world leader since 2013 (IFR, 2019).

In the following paragraphs, three endemic examples of smart robotic setups for manufacturing (CPS) are presented (Wang, Törngren and Onori, 2015). They are classified in three topics: (i) Service-oriented Architecture, (ii) adaptive manufacturing and (iii) model-driven manufacturing systems.

One of the earliest Service-oriented Architectures that was successfully deployed in industry and is still on duty today was fulfilled by the Ford Motor Company on the Valencia assembly factory. In Service-oriented Architectures, the main goal is to align the business layer information flow with the technology specific information flow (Borangiu *et al.*, 2015). Designed by a system assembler (IntRoSys SA), this programme kept all Programmable Logic Controllers (PLCs) controlling the robots as they were. Particular agent technology was implanted as "wrappers" on the PLC according to a structural design built on three diverse agents: (1) a Product Agent (PA) which defines and handles the workflow, (2) a Knowledge Manager

Agent (KMA) that checks the physical viability of the projected workflows and (3) a Machine Agent (MA) that decodes the workflows into exact robot commands. The central feature here is the parameterization of communal procedures, as it is described by the evolvable systems method (Maffei and Onori, 2011). In this regard, agents and PLCs act as extremely flexible CPS.

A second example of autonomous robotic CPS in adaptive manufacturing is the FESTO system, MiniProd (Ferreira, Lohse and Ratchev, 2010; Ribeiro, Barata and Ferreira, 2010). It worked with a multi-agent control arrangement (Agent-Oriented Architecture), could be reprogrammed on the fly and was formed of diverse mechatronic units self-configured by their own implanted controlling systems. The final mechatronic architecture enables the system to be reconfigured for any assembly system layout.

A third example is a 3D model-driven smart robot for remote assembly in the cloud (Wang *et al.*, 2003, 2014), where an off-site worker can operate a robot on the fly by means of an Internet connection. Instead of typical video image streaming, 3D models integrated with kinematics models are utilized to lead the operator during the operation. The 3D models of the entities to be put together are created following a series of snapshots of the parts taken by a camera arranged on the robot at the initial stage. This technology implies a lighter bandwidth consumption than image streaming. For an image frame of 307200 bytes, the equivalent in the 3D model-driven robot, it would consist of only 52 bytes, a significant size reduction eliminating potential bottlenecks (Wang, 2014). In Figure 3.4, a sample case of three actual parts to be assembled is shown. Aided by the CPS, it thus allows real assembly through virtual assembly in real time and lays the groundwork for upcoming factories.

In addition to the previous examples, there are two more actual instances based on collaboration robotics. First, a multiple autonomous robotic collaboration approach for spray painting in which each entity provides a certain capability to the collaborative set (Hassan and Liu, 2017). Second, a series of cooperative assembly manoeuvres which make robot configurations that grab components and make intricate structures such as a chair (Dogar *et al.*, 2018).

Now that general examples of manufacturing applications have been shown, it is time to focus on individual robots. Next, a list of several actual robots with the latest technology which unveil the robotic potential in Industry 4.0 (Bahrin *et al.*, 2016) is presented:

- LBR iiwa (Kuka). This robot is specially meant for safe hand-in-hand human–robot operation and high-precision-required tasks. Connected to the cloud, it is able to make decisions, document its work, and understand on its own.
- Apas family (Bosch). The Apas family robot counts on a dialogue-controlled user interface that enables it to be easily configured on the fly. In addition, its state-of-the-art collision avoidance system places it as a safe solution.
- Nextstage robots (Kawada Industries). Nextstage robots feature an advanced stereo vision system that can accurately get 3D coordinates of a body. Its main application is in part assembly chains.

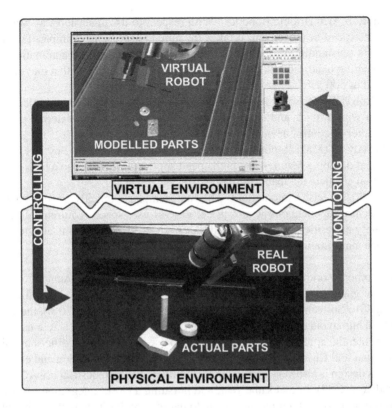

FIGURE 3.4   Remote assemble as a 3D model-driven CPS. Reprinted with permission from Elsevier (Wang, Törngren and Onori, 2015).

- YuMi (ABB). The YuMi robot has high-precision force sensors on the base of high-tech control algorithms in order to programme it through teaching. It is also provided with an integrated safety solution that makes it suitable to work with humans.
- Baxter (Rethink Robotics). This interactive robot is especially suited to production tasks, namely industrial packaging and carrying.
- Sawyer (Rethink Robotics). The Sawyer robot, which updates the previous Baxter model, is a one hand high-performance CoBot designed to carry out duties difficult to run by traditional robots.
- Roberta (Gomtec). This is an economical alternative focusing on small industries that search for automation and flexibility.
- UR series (Universal Robots). Universal Robots created this series in order to improve low-weight cooperative jobs (i.e. carrying, arranging and checking).
- CR-35Ia (Fanuc). This CoBot is able to transport heavier loads, up to 35 kg, which empowers it to work in undesirable conditions. Pursuing a closer interaction with humans, even its colour was chosen to enhance its perception by workers.

- BioRob Arm (Bionic Robotics). With respect to its dimensions, as well as to its speed of movement, it takes the human arm as an example. Thanks to its outstanding properties, BioRob is specialized to the automation of changing handling tasks with small and medium lot sizes with a maximum load of 500 g.
- P-Rob (F&P Personal Robotics). The P-Robot is an all-in-one robotic solution with a quick and simple setup: it doesn't require any external power adapter or control device.
- Speedy-10 (Mabi Robotic). A lightweight robot characterized by its exceptional damping features and high-speed applications thanks to a high-resolution absolute feedback encoder.
- PF400 (Precise Automation). This autosampler explicitly developed for benchtop applications where price, ease-of-use, space requirements and safety are critical. Combined with its space saving design, it is able to service many stations in an extremely small work cell.

Among other worldwide known companies, Foxconn labour stands out. The Taiwanese giant, known for being main manufacturer of Apple products, first fabricated in 2016 Softbank's Pepper humanoid robots series (Russell, 2016), the world's first social humanoid robot able to recognize faces and basic human emotions. Within all their potential applications, it has been put to use in government offices in Taiwan (Chih-hsuan and Chen, 2016). Despite being best suited for education and entertainment, this design is also very promising for commercial and industrial uses (Gardecki and Podpora, 2017). In that same year, and pursuing a closer integration to Industry 4.0, Foxconn presented a three-stepped automation plan for its manufacturing plants (Statt, 2016): first, a replacement of the work that is either dangerous or involves repetitive labour; second, an improvement of the efficiency by streamlining production lines to reduce the number of excess robots in use; and third, the automation of entire factories, "with only a minimal number of workers [...]". In this way, a benchmark of 30 percent automation at its Chinese factories by 2020 was set (Huifeng, 2015).

Another outstanding company, Airbus, the French giant in the field of aeronautics, has done pioneering work in the approach of aircraft manufacturing to I4.0. As it is widely known, this specific sector is still struggling to fully seize the opportunities of I4.0 (Guyon *et al.*, 2019). Strict regulations concerning security make it difficult to change processes, as any change in the manufacturing course means total obedience to the compulsory regulations. Furthermore, the low bulk of production and large amount of diverse tasks to perform makes fixed base robotics an inefficient option in this industry (Guyon *et al.*, 2019). These issues have motivated Airbus to work with JRL (Joint Robotics Laboratory) on a new generation of humanoid robots with enhanced mobility, stability and the capability to shift between different levels on the assembly line (CNRS, 2016). To do that, they were designed so that their entire body connects with the environment, and not only through its feet. Their main tasks are cutting nuts, cleaning, inserting components and even certain duties in quality control. The aircraft manufacturer also worked with Texas Instruments on a set of smart tools focused on basic fastening chores in aircraft assembly (a dangerous stage for workers with risk of human error) (National Instruments, 2015).

3.4   FUTURE SCENARIOS

The current main barrier in manufacturing for a full exploitation of robots (CPS) is the fact that this conservative industry operates under incredibly tight margins and tends to avoid uncertainties (Wang, 2014). Therefore, far from abrupt changes, transition technology is to be established (i.e. the CODESYS solution by the German software company 3S-Smart Software Solutions).

Uncertainty in opportunities and threats is a major concern for future scenarios of robotics and I4.0 (Magruk, 2016). According to some authors, uncertainty appears when economic, political, social, environmental and technical changes take place (Wawiernia, 2013). Nowadays, there are three phenomena which are attributed to broad fields of ignorance about I4.0 (Paprocki, 2016): (i) universal digitalization and ensuring constant communication, (ii) increasingly implemented disruptive technological innovations and (iii) autonomous behaviour of machines using AI.

In order to cope with uncertainty in I4.0, some authors propose various guidelines (Magruk, 2016):

- Adoption of common standards.
- Guarantee of data security through new systems.
- Acknowledgement that human beings will continue to play an active, engaging role in manufacturing. In this way, if machines do not accept human qualities, we will be forced to take over their modus operandi.
- Adaptation of regulation and legislation to new I4.0.
- Create an infrastructure for promoting entrepreneurship.
- I4.0 competing fields embedded in education policies (i.e. programming, data analysis, etc.).

To ensure a positive development, several current organizations have considerable investigation subdivisions focused on support product development and computerized processing through robotics, while also advertising their expertise to other industrial companies and offering platforms for "smart" facilities. This can further motivate the establishment of Industry 4.0. In order to avoid proprietary stand-alone solutions, and so to set up a standardized and open ecosystem to facilitate the operation of factories that work seamlessly with automated logistics and service ecosystems, the Open Industry 4.0 Alliance was created (Hannover Messe, 2019). Next, some noteworthy examples of organizations interested in the robotic industry of I4.0 are presented (Bahrin *et al.*, 2016):

- Industry-Science Research Alliance. This advisory group helps to provide a forum where leaders from the business and science communities discuss topics beyond the boundary of their enterprises. The implementation of the German Federal Government's High-Tech Strategy is the principal aim.
- Acatech. Acatech, established as the German Academy of Science and Engineering, focuses on advising from neutral, qualified and science-based assessments of technology-related matters. In other words, knowledge transference between science and industry is supported.

- SmartFactoryKL. SmartFactoryKL is another pioneering company in respect to transferring technology aspects and putting them into practice.
- German Research Centre for AI (DFKI). The vision of DFKI is based on CPS, combining mechanical systems with electronics to connect everything together where the different modules in the factory could potentially drive themselves around to allow factories to alter the production line.
- Institute for Management Cybernetics, RWTH Aachen University. This institute provides answers and solutions for economic and technological concerns in interdisciplinary teams. Management Cybernetics is defined as a practice which regards businesses and enterprises as open, socio-technological, economic, multifaceted and networked systems, and whose main aim is to describe the complex phenomena that take place within organizations. To this end, models are used with the aid of engineering, economic and social sciences.
- Plattform Industrie 4.0. The Plattform Industrie 4.0 is Germany's central network for driving forward digitalization in manufacturing, trying to understand trends, connecting people and offering support.
- Industrial Internet Consortium (IIC). The IIC brings together the organizations and technologies required to deploy the growth of the Industrial Internet by identifying, assessing and encouraging better practices. Members work together to increase the commercial usage of forward-looking technologies.
- Siemens. In their PLCs manufacturing line they have created a real smart factory, with a state-of-the-art integration on the cloud of automation routines, product logistics, etc.
- W3C. The World Wide Web Consortium is an international society where Member organizations, a full-time staff and the public work together to develop open standards to ensure the long-term progress of the web.
- OPC Foundation. The OPC Foundation defines information as the key to business success and profitability. For this reason, this Foundation's objective is to allow information to be easily and securely exchanged, building a seamless network above a wide variety of data resources.
- Fraunhofer IAO. The Fraunhofer Institute for Labour Economics and Organisation has since 2011 developed strategies, business models and solutions for digital transformation together with companies and institutions in the public sector.
- Google. Google Glass are web-enabled glasses which display dialogue and error messages on a head-up display directly into a person's field of vision. In this way, the operator can take manual action in a safer way as both hands are free.
- Intel and Telit. Intel is a key agent in new generation energy-consuming chips development for IoT connected devices. Leading IoT solution provider Telit is offering deviceWISE, an application enablement platform running on Intel IoT Gateway Technology that greatly simplifies the process of implementing IoT solutions.

- Microsoft. In 2015, Microsoft and Kuka AG presented Kuka LBR iiwa, and Intelligent Industrial Work Assistant built with Microsoft Azure's Internet of Things (IoT) services.
- IBM. IBM focuses on five growing mainstreams – Cloud, BD and Analytics, Mobile, Social Business and Security.
- Bosch Rexroth. Bosch Rexroth provides economic products to build actual assembly lines, such as interfaces to programme PLCs over the air (Open Corte Engineering), interfaces for hydraulic system's controlling (IAC-Multi-Ethernet), motion sensors (HCM model), etc.

In addition to the previous companies, universities are heading the transition to I4.0. To this end, they hold courses which offer manufacturing executives and frontline leaders strategies to implement the technological change at their companies. For the sake of example, the MIT in Cambridge, Massachusetts, is holding a programme in 2020 to introduce new business models and forms of operations based on IoT (MIT, 2019). Depending on cultural and workforce factors, these models would develop a company culture that puts people first. In the Delft University of Technology in the Netherlands, thesis works state that the robot technology is still in an infant step although it has improved immensely in the last ten years, and that maybe in 20 to 25 years robots will be mature enough to develop natural human behaviours with a full understanding of their surroundings and with a suitable human-machine collaboration (Jimenez, 2017).

A controversial matter regarding transition to I4.0 is the role that human beings will play (Nam, 2019). Artificial Intelligence and robots are considered as potential disruptors for work and skills (Rhisiart *et al.*, 2017). Since automation of mass production using robots provides high productivity and robustness, it is commonly thought that human work will progressively vanish (Frey and Osborne, 2017). However, it is ascertained that human expertise may be even more essential than ever (Pfeiffer, 2016). An actual case study in car body construction plant of OEM (Original Equipment Manufacturers) in Germany revealed that 90% of the staff in the assembly line were qualified workers (Pfeiffer, 2016). In fact, they went through three-year vocational training not only to be able to cope with technical incidents of the machinery, but also in order to prevent them. Hence, their 20 to 30 interventions per shift were indispensable and, undoubtedly, non-routine. The research results showed how complex and how dependent on human work environments are, especially on their experiential abilities, which could be seen as the opposite of simple routine. Examples like this make clear that human work, despite the quantitative decrease, is bound to a qualitative increase due to the fact that routine and monotonous tasks would rely on robots.

From this reasoning, there are four aspects where robots will always find limits to their applicability (Pfeiffer, 2016): (i) coping with imponderables and subjective matters, (ii) the flexibility to adapt, (iii) the unrestrained changeability of behaviour and (iv) the skill and body intelligence. Additionally, the potentially negative impact of automation on unskilled labour has spurred the debate on robot taxing (Gasteiger and Prettner, 2017; Shiller, 2017; Zhang, 2019). As an irony of automation, "the more we depend on technology and push it to its limits, the more we need highly skilled, well trained, well-practiced people to make systems resilient, acting as the last time of deference against the failures that will inevitably occur" (Baxter *et al.*, 2012).

Despite uncertainties for the level of automation that the industry will reach in the coming future, data from past years and some projections can help understand what future firms look like. In this regard, the 2019 report by the International Federation of Robots (IFR, 2019) highlights a growing trend since 2010 toward automation and continued technical innovations in industrial robots. This trend was especially important in Asia and Australia, where the number of installed industrial robots increased from 70000 to 283000 units. The number of installed units is expected to rise until 2022, with a forecast of 583520 units in 2022 from the 422271 installed in 2018.

3.5   CONCLUSIONS

The current Industry 4.0 is strongly focused on automation and robotization. Particularly, when attending to manufacturing sectors, these solutions are being used for ending a wide variety of routine and monotonous tasks. In this context, the development of robotic and communication technologies has driven the apparition of a wide range of new applications for robotic solutions, especially those in which human workers and robots increase their interaction.

The present chapter reviewed the current state of autonomous and collaborative robot applications in manufacturing. Initially, a detailed description of robotic technologies and their relation to other key elements of Industry 4.0, including its historical background, was presented. Then, the chapter identified a wide range of applications in manufacturing, mainly highlighting the current situation for the applications of industrial solutions by some of the main worldwide industrial companies (e.g. ABB, Kuka, Universal Robots). Finally, the potential for industrial robots was discussed based on current data on installation of robotic solutions and projections, along with the identification of some of the main actors (e.g. Google, Microsoft, Siemens) with potential to boost the adoption of these solutions under the Industry 4.0 framework in the coming years.

REFERENCES

Alcácer, V. and Cruz-Machado, V. (2019) 'Scanning the industry 4.0: A literature review on technologies for manufacturing systems', *Engineering Science and Technology, an International Journal*, 22(3), pp. 899–919. Elsevier. doi:10.1016/J.JESTCH.2019.01.006.

Ariyaluran, R. A. *et al.* (2018) 'Real-time big data processing for anomaly detection: A Survey', *International Journal of Information Management*. doi:10.1016/j.ijinfomgt.2018.08.006.

Baheti, R. and Gill, H. (2011) 'Cyber-physical systems', *The Impact of Control Technology*, IEEE Control Systems Society, pp. 161–166.

Bahrin, M. A. K. *et al.* (2016) 'Industry 4.0: A review on industrial automation and robotic', *Jurnal Teknologi*, 78(6–13). doi:10.11113/jt.v78.9285.

Balint, T. S. and Hall, A. (2015) 'Humanly space objects—Perception and connectionwith the observer', *Acta Astronautica*, 110, pp. 129–144. doi:10.1016/j.actaastro.2015.01.010.

Baxter, G. *et al.* (2012) 'The ironies of automation... still going strong at 30?' *ACM International Conference Proceeding Series*. doi:10.1145/2448136.2448149.

Ben-Ari, M. and Mondada, F. (2017) 'Robots and their applications', In *Elements of Robotics*. Cham: Springer International Publishing, pp. 1–20. doi:10.1007/978-3-319-62533-1_1.

Blanchet, M. *et al*. (2014) 'Industry 4.0 - The new industrial revolution — How Europe will succed'. Roland Berger Strategy Consultants. Available at: www.rolandberger.com/en/P ublications/Industry-4.0--the-new-industrial-revolution.html (Accessed: 8 August 2019).

Borangiu, T. *et al*. (2015) *A Service Oriented Architecture for Total Manufacturing Enterprise Integration*. doi:10.1007/978-3-319-14980-6_8.

Brending, S. *et al*. (2017) 'Certifiable software architecture for human robot collaboration in industrial production environments', *IFAC-PapersOnLine*, 50(1), pp. 1983–1990. doi:10.1016/j.ifacol.2017.08.171.

Brodsky, A. *et al*. (2016) 'Analysis and optimization based on reusable knowledge base of process performance models', *The International Journal of Advanced Manufacturing Technology*, 88(1–4), pp. 337–357. London: Springer. doi:10.1007/s00170-016-8761-7.

Cengarle, V. *et al*. (2013) *Structuring of CPS Domain: Characteristics, Trends, Challenges and Opportunities Associated with CPS*. Available at: http://www.cyphers.eu/sites/ default/files/D2.2.pdf.

Chih-hsuan, K. and Chen, C. (2016) 'Pepper the robot to work at Pingtung County government', *FOCUS TAIWAN - CNA ENGLISH NEWS*. Available at: http://focustaiwan.tw/ news/ast/201607220017.aspx (Accessed: 16 August 2019).

CNRS. (2016) *Des robots humanoïdes dans les usines aéronautiques de demain*. Paris, France. https://www.techniques-ingenieur.fr/actualite/articles/des-robots-humanoide s-dans-les-usines-aeronautiques-de-demain-32042/

Colombo, A. W., Schoop, R. and Neubert, R. (2006) 'An agent-based intelligent control platform for industrial holonic manufacturing systems', *IEEE Transactions on Industrial Electronics*, 53(1), pp. 322–337. doi:10.1109/TIE.2005.862210.

Corrales, J. A., Candelas, F. A. and Torres, F. (2011) 'Safe human–robot interaction based on dynamic sphere-swept line bounding volumes', *Robotics and Computer-Integrated Manufacturing*, 27(1), pp. 177–185. Pergamon. doi:10.1016/J.RCIM.2010.07.005.

Dalenogare, L. S. *et al*. (2018) 'The expected contribution of Industry 4.0 technologies for industrial performance', *International Journal of Production Economics*, 204, pp. 383–394. Elsevier. doi:10.1016/J.IJPE.2018.08.019.

Davis, J. *et al*. (2015) 'Smart Manufacturing', *Annual Review of Chemical and Biomolecular Engineering*, 6(1), pp. 141–160. Annual Reviews. doi:10.1146/ annurev-chembioeng-061114-123255.

Deese, J. M. *et al*. (2018) 'Patient reported and clinical outcomes of robotic-arm assisted unicondylarknee arthroplasty: Minimum two year follow-up', *Journal of Orthopaedics*, 15(3), pp. 847–853. doi:10.1016/j.jor.2018.08.018.

DIN/DKE. (2018) *German Standardization Roadmap. Industry 4.0*. Version 3 Available at: https://www.din.de/resource/blob/65354/57218767bd6da1927b181b9f2a0d5b39/road-map-i4-0-e-data.pdf (Accessed: 22 February 2020).

Dogar, M. *et al*. (2018) 'Multi-robot grasp planning for sequential assembly operations', *Autonomous Robots*, 43(3), pp. 649–664. Springer US. doi:10.1007/s10514-018-9748-z.

Elisabetta, R. (2016) 'Smart work', *Evidence-based HRM: A Global Forum for Empirical Scholarship*. In G. Luca (Ed.). Vol. 4(3). Emerald Group Publishing Limited, pp. 240–256. doi:10.1108/EBHRM-01-2016-0004.

Ferreira, P., Lohse, N. and Ratchev, S. (2010) 'Multi-agent architecture for reconfiguration of precision modular assembly systems', Springer, Berlin, Heidelberg, pp. 247–254. doi:10.1007/978-3-642-11598-1_29.

Fischer, M. and Henrich, D. (2009) '3D collision detection for industrial robots and unknown obstacles using multiple depth images', In *Advances in Robotics Research*. Berlin, Heidelberg: Springer Berlin Heidelberg, pp. 111–122. doi:10.1007/978-3-642-01213-6_11.

Flacco, F. *et al.* (2012) 'A depth space approach to human-robot collision avoidance', In *2012 IEEE International Conference on Robotics and Automation.* IEEE, pp. 338–345. doi:10.1109/ICRA.2012.6225245.

Frank, A. G., Dalenogare, L. S. and Ayala, N. F. (2019) 'Industry 4.0 technologies: Implementation patterns in manufacturing companies', *International Journal of Production Economics*, 210, pp. 15–26. Elsevier. doi:10.1016/J.IJPE.2019.01.004.

Frey, C. B. and Osborne, M. A. (2017) 'The future of employment: How susceptible are jobs to computerisation?' *Technological Forecasting and Social Change*, 114, pp. 254–280 doi:10.1016/j.techfore.2016.08.019.

Gardecki, A. and Podpora, M. (2017) 'Experience from the operation of the Pepper humanoid robots', In *2017 Progress in Applied Electrical Engineering (PAEE).* IEEE, pp. 1–6. doi:10.1109/PAEE.2017.8008994.

Gasteiger, E. and Prettner, K. (2017) 'A note on automation, stagnation, and the implications of a robot tax', *Discussion Papers.* Free University Berlin, School of Business & Economics. Available at: https://ideas.repec.org/p/zbw/fubsbe/201717.html (Accessed: 16 August 2019).

Gerbert, P. *et al.* (2015) 'Industry 4.0: The future of productivity and growth in manufacturing industries'. Available at: https://www.bcg.com/publications/2015/engineered_p roducts_project_business_industry_4_future_productivity_growth_manufacturing_ industries.aspx (Accessed: 8 August 2019).

Guyon, I. *et al.* (2019) *Analysis of the Opportunities of Industry 4.0 in the Aeronautical Sector.* Available at: https://hal.archives-ouvertes.fr/hal-02063948 (Accessed: 17 August 2019).

Hannover Messe. (2019) *Large Enterprises Set Up an Alliance for Industry 4.0.* Available at: https://www.hannovermesse.de/en/news/news-overview/large-enterprises-set-up-an-a lliance-for-industry-4.0-121408.xhtml (Accessed: 17 August 2019).

Hassan, M. and Liu, D. (2017) 'Simultaneous area partitioning and allocation for complete coverage by multiple autonomous industrial robots', *Autonomous Robots*, 41(8), pp. 1609–1628. Springer US. doi:10.1007/s10514-017-9631-3.

Ho, J.-Y. and O'Sullivan, E. (2017) 'Strategic standardisation of smart systems: A roadmapping process in support of innovation', *Technological Forecasting and Social Change*, 115, pp. 301–312. North-Holland. doi:10.1016/J.TECHFORE.2016.04.014.

Huifeng, H. (2015) 'Foxconn's Foxbot army close to hitting the Chinese market, on track to meet 30 per cent automation target', *South China Morning Post.* Available at: https ://www.scmp.com/tech/innovation/article/1829834/foxconns-foxbot-army-close-hitti ng-chinese-market-track-meet-30-cent (Accessed: 16 August 2019).

IFR (International Federation for Robotics). (2019) 'IFR. World Robotics. Industrial Robots 2019' (Executive summary). https://ifr.org/downloads/press2018/Executive%20Su mmary%20WR%202019%20Industrial%20Robots.pdf

Jimenez, M. (2017) *What is the Impact of the Industry 4.0 in the Process Industry?* Delft University of Technology. Available at: http://resolver.tudelft.nl/uuid:c4a8ac12-2b71 -4780-8d01-55d6a9bed6ca (Accessed: 18 August 2019).

Kagermann, H. (2013) *Recommendations for Implementing the Strategic Initiative INDUSTRIE 4.0 – Securing the Future of German Manufacturing Industry | BibSonomy.* Available at: http://www.acatech.de/fileadmin/userupload/Baumstr ukturnachWebsite/Acatech/root/de/Material fuerSonderseiten/Industrie (Accessed: 3 August 2019).

Klocke, F. *et al.* (2011) 'Process monitoring and control of machining operations', *International Journal of Automation Technology*, 5(3), pp. 403–411. Fuji Technology Press Ltd. doi:10.20965/ijat.2011.p0403.

Koch, P. J. *et al.* (2017) 'A skill-based robot co-worker for industrial maintenance tasks', *Procedia Manufacturing*, 11, pp. 83–90. Elsevier. doi:10.1016/J.PROMFG.2017.07.141.

Krüger, J. *et al.* (2005) 'Image based 3D Surveillance for flexible Man-Robot-Cooperation', *CIRP Annals*, 54(1), pp. 19–22. Elsevier. doi:10.1016/S0007-8506(07)60040-7.

Lee, E. A. (2006) 'Cyber-physical systems - are computing foundations adequate?' *Position Paper for NSF Workshop on Cyber-Physical Systems: Research Motivation, Techniques and Roadmap.*

Liu, F. *et al.* (2000) 'Real-time systems'. Available at: http://citeseerx.ist.psu.edu/viewdoc/summary?doi=10.1.1.387.1414 (Accessed: 13 August 2019).

Liu, H. and Wang, L. (2018) 'Gesture recognition for human-robot collaboration: A review', *International Journal of Industrial Ergonomics*, 68, pp. 355–367. Elsevier. doi:10.1016/J.ERGON.2017.02.004.

Longo, F., Nicoletti, L. and Padovano, A. (2017) 'Smart operators in industry 4.0: A human-centered approach to enhance operators' capabilities and competencies within the new smart factory context', *Computers & Industrial Engineering*, 113, pp. 144–159. Pergamon. doi:10.1016/J.CIE.2017.09.016.

Maffei, A. and Onori, M. (2011) 'Evolvable production systems: Environment for new business models', *Key Engineering Materials*, 467–469, pp. 1592–1597. Trans Tech Publications Ltd. doi:10.4028/www.scientific.net/KEM.467-469.1592.

Magruk, A. (2016) 'Uncertainty in the sphere of the industry 4.0 – potential areas to research', *Business, Management and Education*, 14(2), pp. 275–291. doi:10.3846/bme.2016.332.

El Makrini, I. *et al.* (2018) 'Working with Walt: How a Cobot was developed and inserted on an auto assembly line', *IEEE Robotics & Automation Magazine*, 25(2), pp. 51–58. doi:10.1109/MRA.2018.2815947.

Maturana, F. P. and Norrie, D. H. (1996) 'Multi-agent mediator architecture for distributed manufacturing', *Journal of Intelligent Manufacturing*, 7(4), pp. 257–270. Kluwer Academic Publishers. doi:10.1007/BF00124828.

MIT. (2019) 'Implementing industry 4.0: Leading change in manufacturing and operations'. Cambridge, MA. Available at: https://executive.mit.edu/openenrollment/program/implementing-industry-40-leading-change-in-manufacturing-operations/#program-details (Accessed: 17 August 2019).

Mohammed, W. M. *et al.* (2018) 'Generic platform for manufacturing execution system functions in knowledge-driven manufacturing systems', *International Journal of Computer Integrated Manufacturing*, 31(3), pp. 262–274. Taylor & Francis. doi:10.1080/0951192X.2017.1407874.

Monostori, L., Váncza, J. and Kumara, S. R. T. (2006) 'Agent-based systems for manufacturing', *CIRP Annals*, 55(2), pp. 697–720. Elsevier. doi:10.1016/J.CIRP.2006.10.004.

Nam, T. (2019) 'Citizen attitudes about job replacement by robotic automation', *Futures*, 109, pp. 39–49.

National Instruments. (2015) *Developing Smart Tools for the Airbus Factory of the Future.* Needham, MA: Industrial Internet Consortium & National Instruments.

Nord, J. H., Koohang, A. and Paliszkiewicz, J. (2019) 'The internet of things: Review and theoretical framework', *Expert Systems with Applications*, 133, pp. 97–108. Pergamon. doi:10.1016/J.ESWA.2019.05.014.

Nye, D. E. (2013) *America's Assembly Line.* The MIT Press. Available at: https://www.jstor.org/stable/j.ctt5vjr2k (Accessed: 11 August 2019).

Papazoglou, M. P., van den Heuvel, W.-J. and Mascolo, J. E. (2015) 'A reference architecture and knowledge-based structures for smart manufacturing networks', *IEEE Software*, 32(3), pp. 61–69. doi:10.1109/MS.2015.57.

Paprocki, W. (2016). 'Koncepcja Przemysł 4.0 i jej zastosowanie w warunkach gospodarki cyfrowej', In J. Gajewski, W. Paprocki, J. Pieriegud (Eds.). *Cyfryzacja gospodarki i społeczeństwa. Szanse i wyzwania dla sektorów infrastrukturalnych.* Gdańsk: Publikacja Europejskiego Kongresu, Finansowego, pp. 39–57.

Pedersen, M. R. *et al.* (2016) 'Robot skills for manufacturing: From concept to industrial deployment', *Robotics and Computer-Integrated Manufacturing*, 37, pp. 282–291. Pergamon. doi:10.1016/J.RCIM.2015.04.002.

Pfeiffer, S. (2016) 'Robots, industry 4.0 and humans, or why assembly work is more than routine work', *Societies*, 6(2), p. 16. Multidisciplinary Digital Publishing Institute. doi:10.3390/soc6020016.

Rehman, M. H. *et al.* (2019) 'The role of big data analytics in industrial Internet of Things', *Future Generation Computer Systems*, 99, pp. 247–259. North-Holland. doi:10.1016/J. FUTURE.2019.04.020.

Rhisiart, M. Störmer, E. Daheim, C. (2017) 'From foresight to impact? The 2030 future of work scenarios', *Technological Forecasting and Social Change*, 124, pp. 203–213. doi:10.1016/j.techfore.2016.11.020.

Ribeiro, L., Barata, J. and Ferreira, J. (2010) 'An agent-based interaction-oriented shop floor to support emergent diagnosis', In *2010 8th IEEE International Conference on Industrial Informatics*. IEEE, pp. 189–194. doi:10.1109/INDIN.2010.5549436.

Rizzi, A. A., Gowdy, J. and Hollis, R. L. (2001) 'Distributed Coordination in Modular Precision Assembly Systems', *The International Journal of Robotics Research*, 20(10), pp. 819–838. SAGE Publications. doi:10.1177/02783640122068128.

Russell, J. (2016) 'SoftBank's Pepper robot goes on sale in Taiwan in first overseas launch', *TechCrunch*. Available at: https://techcrunch.com/2016/07/25/softbanks-pepper-robot-goes-on-sale-in-taiwan-in-first-overseas-launch/?guccounter=1&guce_referrer_us=aHR0cHM6Ly93d3cuYmluZy5jb20v&guce_referrer_cs=_grvzLDJcnhWjby wOeRNPA (Accessed: 16 August 2019).

Salkin, C. *et al.* (2017) 'A conceptual framework for industry 4.0', Springer, Cham, pp. 3–23. doi:10.1007/978-3-319-57870-5_1.

Schiavi, R., Bicchi, A. and Flacco, F. (2009) 'Integration of active and passive compliance control for safe human-robot coexistence', In *2009 IEEE International Conference on Robotics and Automation*. IEEE, pp. 259–264. doi:10.1109/ROBOT.2009.5152571.

Schuh, G. *et al.* (2017) *Industrie 4.0 Maturity Index. Managing the Digital Transformation of Companies*. Available at: https://www.acatech.de/wp-content/uploads/2018/03/acate ch_STUDIE_Maturity_Index_eng_WEB-1.pdf (Accessed: 6 August 2019).

Shiller, R. J. (2017) 'Robotization without taxation?' *Project Syndicate*. Available at: https ://www.project-syndicate.org/commentary/temporary-robot-tax-finances-adjustment-by-robert-j--shiller-2017-03?barrier=accesspaylog (Accessed: 16 August 2019).

Shirase, K. and Nakamoto, K. (2013) 'Simulation technologies for the development of an autonomous and intelligent machine tool', *International Journal of Automation Technology*, 7(1), pp. 6–15. Fuji Technology Press Ltd. doi:10.20965/ijat.2013.p0006.

SMLC. (2011) *Implementing 21st Century Smart Manufacturing*. Washington D.C. Available at: https://www.controlglobal.com/whitepapers/2011/110621-smlc-smart-manufactur ing/ (Accessed: 5 August 2019).

de Sousa Jabbour, A. B. L. *et al.* (2018) 'When titans meet – Can industry 4.0 revolutionise the environmentally-sustainable manufacturing wave? The role of critical success factors', *Technological Forecasting and Social Change*, 132, pp. 18–25. North-Holland. doi:10.1016/J.TECHFORE.2018.01.017.

Statt, N. (2016) *iPhone Manufacturer Foxconn Plans to Replace Almost Every Human Worker with Robots - The Verge*. Available at: https://www.theverge.com/2016/12/30 /14128870/foxconn-robots-automation-apple-iphone-china-manufacturing (Accessed: 16 August 2019).

Stock, T. *et al.* (2018) 'Industry 4.0 as enabler for a sustainable development: A qualitative assessment of its ecological and social potential', *Process Safety and Environmental Protection*, 118, pp. 254–267. Elsevier. doi:10.1016/J.PSEP.2018.06.026.

Surdilovic, D., Schreck, G. and Schmidt, U. (2010) 'Development of collaborative robots (COBOTS) for flexible human-integrated assembly automation', *undefined*. Available at: https://www.semanticscholar.org/paper/Development-of-Collaborative-Robots-(COBOTS)-for-Surdilovic-Schreck/1458cc78d07a3f219f9a0b6984b75e5b16de6dcd (Accessed: 5 August 2019).

Tao, F. *et al.* (2018) 'Data-driven smart manufacturing', *Journal of Manufacturing Systems*, 48, pp. 157–169. doi:10.1016/j.jmsy.2018.01.006.

Thoben, K.-D., Wiesner, S. and Wuest, T. (2017) '"Industrie 4.0" and smart manufacturing – a review of research issues and application examples', *International Journal of Automation Technology*, 11(1), pp. 4–16. doi:10.20965/ijat.2017.p0004.

Wang, L. *et al.* (2003) 'Integrating Java 3D model and sensor data for remote monitoring and control', *Robotics and Computer-Integrated Manufacturing*, 19(1–2), pp. 13–19. Pergamon. doi:10.1016/S0736-5845(02)00058-3.

Wang, L. (2014) *Cyber Manufacturing: Research and Applications*. Proceedings of the 10th International Symposium on Tools and Methods of Competitive Engineering, pp. 39–49. Budapest, Hungary, May 19–23.

Wang, L. *et al.* (2014) 'A cloud-based approach for WEEE remanufacturing', *CIRP Annals*, 63(1), pp. 409–412. Elsevier. doi:10.1016/J.CIRP.2014.03.114.

Wang, L. *et al.* (2018) 'Energy-efficient robot applications towards sustainable manufacturing', *International Journal of Computer Integrated Manufacturing*, 31(8), pp. 692–700. Taylor & Francis. doi:10.1080/0951192X.2017.1379099.

Wang, L., Törngren, M. and Onori, M. (2015) 'Current status and advancement of cyberphysical systems in manufacturing', *Journal of Manufacturing Systems*, 37, pp. 517–527. Elsevier. doi:10.1016/J.JMSY.2015.04.008.

Wang, S. *et al.* (2016) 'Implementing smart factory of industrie 4.0: An outlook', *International Journal of Distributed Sensor Networks*, 12(1), p. 3159805. SAGE PublicationsSage UK: London, England. doi:10.1155/2016/3159805.

Wawiernia, A. (2013) *Taksonomia niepewności, Zarządzanie i Finanse*. Wydział Zarządzania - Uniwersytet Gdański. Available at: http://yadda.icm.edu.pl/yadda/element/bwmeta1 .element.desklight-abc5751d-2ad3-4721-81af-a3f3e9beb9f9 (Accessed: 10 August 2019).

Weiss, A. and Huber, A. (2016) 'User experience of a smart factory robot: Assembly line workers demand adaptive robots'. Available at: http://arxiv.org/abs/1606.03846 (Accessed: 5 August 2019).

Wu, Q., Liu, Y. and Wu, C. (2018) 'An overview of current situations of robot industry development', In K. Eguchi and T. Chen (Eds.). *ITM Web of Conferences*. vol. 17. EDP Sciences, p. 03019. doi:10.1051/itmconf/20181703019.

Xu, L. Da, Xu, E. L. and Li, L. (2018) 'Industry 4.0: State of the art and future trends', *International Journal of Production Research*, 56(8), pp. 2941–2962. Taylor & Francis. doi:10.1080/00207543.2018.1444806.

Zhang, J. *et al.* (2017) *Self-Organizing Manufacturing: Current Status and Prospect for Industry 4.0*. doi:10.1109/ES.2017.59.

Zhang, P. (2019) 'Automation, wage inequality and implications of a robot tax', *International Review of Economics & Finance*, 59, pp. 500–509. JAI. doi:10.1016/J.IREF.2018.10.013.

Zhang, Y. *et al.* (2017) 'A big data analytics architecture for cleaner manufacturing and maintenance processes of complex products', *Journal of Cleaner Production*, 142, pp. 626–641. Elsevier. doi:10.1016/J.JCLEPRO.2016.07.123.

4 Augmented Reality and Virtual Reality

From the Industrial Field to Other Areas

Guido Guizzi, Roberto Revetria, and Anastasiia Rozhok

CONTENTS

4.1 INTRODUCTION

Nowadays many manufacturing companies are beginning to encounter problems such as changes in demand and requirements from customers and suppliers. Industrial companies need innovative technologies in production systems that would target workers, their safety and security, new manufacturing processes, where the human errors would not appear. Given this, technologies such as augmented and virtual realities can be used to train staff, which could significantly increase manufacturing efficiency and help to prevent problems with human injuries.

4.2 DESCRIPTION OF THE TECHNOLOGY IN THE CONTEXT OF INDUSTRY 4.0

In recent years, manufacturing companies have faced various challenges related to ever-changing demands from customers and suppliers. New technological roadmaps and suggested interventions aim to exploit the economic potential resulting from the continuing impact of rapidly advancing Information and Communication Technology

(ICT) in the industry. Today we are witnessing the beginning of the fourth industrial revolution, which is the result of the digital revolution. It is characterised by a widespread application of the Internet, which has become easier to use, and the application of intelligent components, robots, and technologies. This revolution supports the creation of the "smart factory" in which physical and digital systems are integrated to reach "mass personalisation" and "faster product development" (Demartini et al., 2017). This new paradigm is known as the "fourth industrial revolution" or "Industry 4.0". In general, when we talk about Industry 4.0, we refer to the way that products are produced, thanks to the digitisation of manufacturing. This transition is so compelling that it is being called Industry 4.0 to represent the fourth revolution that has occurred in manufacturing (Kolberg and Zühlk, 2015). Describing this in a more analytical way, the term refers to a new phase in the industrial revolution that focuses heavily on interconnectivity automation, machine learning, and real-time data.

The innovation strategy of this industrial revolution is not directed towards workerless production (as in the Computer Integrated Manufacturing approach of the 1980s). The human operator is considered to be the most flexible element in the production system because they are the ones who have the highest capacity to adapt (Jimeno and Puerta, 2007). Facing the problem of increasingly complex products, short product development cycles, and fast changes in production processes it will be expected more and more from the worker. Thus, virtual training is, therefore, a promising solution in this context (Gorecky et al., 2017). Also, AR is used in assembly lines to help operators perform tasks that require a high level of qualification, to learn by working, and to help the operator in the execution of Standard Operating Procedures (SOP). The AR can also play an essential role in the processing phases of the shop floor. The operator can carry out work on the machine and, at the same time, acquire information in real time: the environmental impact of operation, safety conditions in the execution of the work, energy consumption (Mawson and Hughes, 2019).

Moreover, Industry 4.0 is sometimes also referred to as the "Industrial Internet of Things" (IIoT) or "Smart Manufacturing". This is because it technically marries physical production and operations with smart digital technology, machine learning, and big data to create a more holistic and better-connected ecosystem for companies that focus on manufacturing and supply chain management.

The term "Industry 4.0" refers to new production patterns, including new technologies, productive factors, and labour organisations, which are completely changing the production processes and the relationship between customer and company with relevant effects on the supply and value chains (Demartini et al., 2017). Even though most of the innovations mentioned above are in an embryonic stage, they are still an essential part of research and progress. The association of these cause new "matched technologies" which could work in a physical and digital environment. These changes in business models, production paradigm, and logistic operations are driving various production sectors and replacing the traditional industrial systems, bringing in the fourth industrial revolution.

In the Industry 4.0 era, a fundamental role is played by the application of new digital and informatics technologies such as the Augmented and Virtual (A/V) reality, and the Internet of Things. These technologies offer the possibility of implementing innovative systems related to industrial safety and security (Revetria et al., 2019).

Even though most of the innovations mentioned above are in an embryonic stage, they are still an essential part of research and progress. The association of these cause new "matched technologies" which could work in a physical and digital environment offering the possibility of improving industrial safety and security among other topics.

It offers a particular focus on industrial safety, based on new interconnected technologies (Mladenov et al., 2018) which can highlight risks and dangers within a working environment. The technologies adopted for this scope are Augmented Reality (AR), the Industrial Internet of Things (IIoT), digital twin, and simulation. AR is defined as a real-time direct or indirect view of a physical real-world environment that has been enhanced by adding virtual computer-generated information to it. This technology finds an application in the most diverse situations: from the warehouse to the production, up to safety and training (Demartini et al., 2017).

As stated earlier, the Augmented and Virtual (A/V) reality plays a fundamental role in the Industry 4.0 era. Thanks to these information technologies, it is possible to implement information systems related to industrial safety. Therefore, it is worthwhile to understand in more detail the concept of Augmented and Virtual (A/V) reality.

To analyse the idea of Augmented Reality (AR), we must first be able to explain and understand what the term augmented reality represents in the real world. To begin with, augmented reality is far beyond QR-code scanning. What augmented reality really does is use existing reality and physical objects to trigger computer-generated enhancements over the top of reality, in real time. Thus, explaining it more commonly, augmented reality is a technology that lays computer-generated images over a user's view of the real world. Typically, those images take shape as 3D models, videos, and information.

Of course, to achieve all of the above, the user must be equipped with the necessary equipment corresponding to each AR application. Before we get there though, let us explain how Augmented Reality (AR) works.

As mentioned earlier, the first thing needed is the necessary equipment. All those devices (glasses, smartphones, PCs) contain software, sensors, and digital projectors that trigger digital displays onto physical objects. The representation of those physical objects varies in each device. One of the examples is Google glasses which displays 2D images onto see-through glasses, while Microsoft HoloLens embeds 3D images into the world around us. The augmented reality process uses a camera with "scanning mode". Once the camera detects a "triggering" object, a digital object onscreen will appear in the position of the real target object. Sometimes, the user will have to interact with several objects to create a database before being able to get information on the target object. AR glasses, computers, and several other devices can use the augmented reality technology. Yet, all these devices need some basic components such as software, hardware, cameras, and applications.

Thus, we can understand that the representation and the way augmented reality works, depends mostly on the device and its specific characteristics.

Thereby, immersive A/V reality technologies open up the possibility of new ways to interact with the workers, the shop floor, and the whole enterprise. Workers strengthen their perception of the product, and its processes become active.

4.3 HISTORICAL EVOLUTION: ORIGINS AND CURRENT STATE. STRENGTHS AND WEAKNESSES

Augmented Reality (AR) is the direct or indirect view of the real world improved or increased through the addition of virtual information designed on the computer. The AR is usually interactive and created three-dimensionally so that you can harmonise with the real world. In Virtual Reality (VR) – also called Virtual Environment – the person is completely immersed in a simulated world that resembles reality or is detached from it. So, between the Real Environment and the Virtual Environment, we find a continuum of additional layers that represent possible mixes deriving from these extremes. Figure 4.1 shows a representation of the Reality-Virtuality Continuum.

Therefore, the difference between the two situations mentioned above is that in AR the real world prevails (from which the user is not detached) which is superimposed on a so-called digital layer, while VR is the exact reverse. Differently from augmented reality, virtual reality is the creation of digital environments or worlds by computers, where the user can interact directly with this reality as a main or secondary actor.

The union of four variables, also called the four I's of VR, gives the operation of VR as shown in Figure 4.2: interaction, immersion, imagination, and intelligence.

Interaction is the interface used by the user to relate to the virtual world. It involves real-time simulation and interaction through a set of different sensory channels (sight, hearing, touch, smell, and taste). Immersion occurs when the user loses contact with physical reality, perceiving only those that come from the virtual world, which becomes primary stimuli. Imagination is the mental ability to represent images of real or ideal things, and it is also the ability to create, conceive, and design new things. Intelligence is the ability of the VR system to understand the actions that

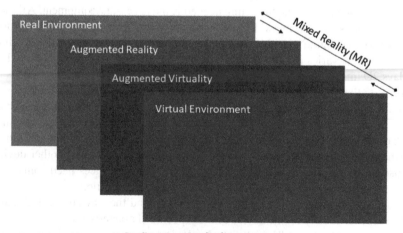

Reality-Virtuality (RV) Continuum

FIGURE 4.1 The reality-virtuality continuum (Based on Milgram et al., 1995).

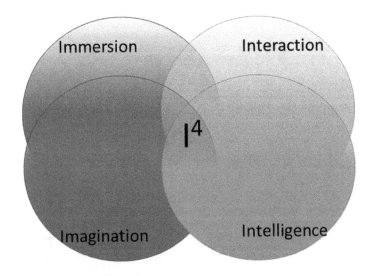

FIGURE 4.2 Quadrilateral of virtual reality (Based on Al-Ahmari et al. 2016).

the user wants to perform (e.g. take an object, use a tool) or understand the situation and propose some things that can be done.

Currently, this type of industry is evolving rapidly along with the development of new applications related to mobile. Using the display of the smartphone, manufacturers have created boxes, i.e. viewers that contain within them the mobile device, which uses special lenses placed a few inches from it, that enhance the vision of the screen, dividing it into two distinct parts thus simulating stereoscopy. The most typical examples of this kind of viewers are Google Cardboard and the Samsung Gear VR.

In the field of AR, many researchers have expressed their vision and the characteristics that this technology should meet. Azuma (Azuma, 1997), for example, does not restrict to a single kind of display (the viewer, for example) the possibility of showing AR and he does not limit even AR to enrich only the sense of sight. Through technology, all five senses can be theoretically "increased": the AR can be used to replace or enhance the missing senses of the user (such as through sight correction or hearing improvement). Virtual objects add information that would otherwise not be directly perceptible to the user. The information sent electronically can help the user in carrying out their daily work activities, such as in the maintenance of electrical cables of an aircraft thanks to an optical device (Azuma, 1997).

The first scientist to propose an augmented reality experience was Morton Heling in the late 1950s, through a machine called Sensorama. This invention aimed to extend the cinematographic experience to all five senses: sounds, smells, wind and oscillations of the machine, to simulate movement, accompanied the proposed short films.

Interest in augmented reality can be assessed by consulting Google's search statistics for the "augmented+reality" key. It can be seen that since 2009, there has been substantial growth, which led it to overcome virtual reality in popularity for the first time. In the graph in Figure 4.3, the value of 100 indicates the highest search frequency of the term in the period considered; 50 indicates half of the searches; the

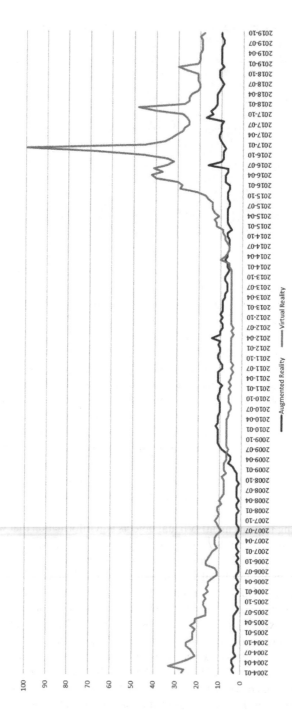

FIGURE 4.3 Google trends: research of "augmented+reality" and "virtual+reality" (Source http://trends.google.it).

value of 0 indicates a search frequency of the term less than 1% compared to the highest search frequency.

The reason for this surge is due to the creation of applications on the desktop via the Web: in fact, the web browser compatible version of the most important framework for the recording of virtual objects within an audiovisual stream (FLARToolkit by ARToolKit) was realised. Besides, the major smartphone manufacturers have invaded the mobile devices market, allowing the development of applications from third parties: the first augmented reality browsers (Aurasma now become HP Reveal, Acrossair and others) appeared.

Interest in augmented reality began to grow when the creatives commissioned by General Electric Company for the new Smart Grid communication campaign created the first application that can be used online. In the first quarter of 2014, it can be seen that interest in VR again overtakes interest in AR in terms of popularity. Probably also because at that time it was expected that the explosion in sales of an AR prototype device such as Google Glass served as a representative of this technology. However, the expected growth did not occur, and therefore the interest in AR decreased. At the same time in 2014, the release of new VR devices, such as Oculus Rift, generated a renewed interest in VR especially in the field of gaming. In 2014 Facebook bought Oculus for US$ 2 billion.

"Imagine sharing not just moments with your friends online, but entire experiences and adventures ... One day, we believe this kind of immersive, augmented reality will become a part of daily life for billions of people" (Mark Zuckerberg, FB post, March 25, 2014).

Today the interest in AR and VR is quite similar because current technologies tend to mainly represent situations of mixed reality.

It is the second-generation augmented reality that is characterised by the current and flourishing season of desktop and mobile applications: from road navigator to videogames. The augmented reality version of the navigators allows the user, who has installed the app on their smartphone, to follow the graphic directions both by displaying them on normal street maps and directly, and interactively, on the video stream of the urban environment.

Over time, manufacturers have developed many devices to implement AR. The devices are divided into holographic devices and immersive devices. As for the first ones, they offer the possibility to insert digital content in the real world and interact with them as if they were there. The most outstanding example is Microsoft HoloLens 2, which is characterised by a display that allows the user to see the physical environment around them while wearing it; it has no cable, so the user has full freedom of movement.

Another holographic device is the Magic Leap. These devices will make the holograms more defined, as the user will have the perception that they are real.

As for the second type of device, that is, the immersive ones, we can find the Project Tango + VR Head-Mounted Device. This Lenovo device initially scans the space in which it interacts, which will be inserted into the viewer. The user will, therefore, be immersed in a virtual space, which has the same limits as the real space that surrounds it, so even if wholly isolated from reality, you will have the opportunity to move freely in a given environment.

Another example is provided by the Acer Mixed Reality that, unlike the first, does not need to be inserted inside a mobile device to work, but it needs to be connected to a PC. The device is equipped with two frontal cameras that capture the elements of reality to insert them into the virtual experience.

AR and VR technologies have strengths and weaknesses.

In particular, the main strengths of AR are:

- The rapid evolution of technologies both from a hardware and a software point of view. These two aspects are strongly interconnected. Some features that you cannot get with the hardware devices are obtained by properly developing the software and vice versa. Moreover, the intuitiveness of the software interfaces represents an essential characteristic of the diffusion of technology.
- The high number of suppliers of components for the production of devices. This aspect is crucial because it should allow, over time, an improvement in the quality of the components and a reduction in their cost.
  The main weaknesses of AR are:
- Still very high prices of the viewers to be able to be spread widely. Current viewers cost between US$2,500 and US$4,000.
- Reputation problems of smart glasses due to the failure of Google Glass. The market did not accept them to the extent that Google expected such futuristic glasses to achieve in 2012. In fact, except for specific applications, people did not believe that, from a functional point of view, the new possibilities offered by such a device were appropriate to its price.
- Health. In general, it is advisable not to use such devices for too long.

For VR, the strengths and weaknesses of this technology have many elements of analogy with AR. VR has great potential: in the future, when it has spread into the market, it can act as an encyclopedia of experiences. For example, at any time you can experience the moments when a volcano erupts, visit archaeological sites, visit Mars, share your stories, view news reports, try non-human experiences such as living like an animal, play increasingly realistic video games, or live the experience of feeling and seeing your heartbeat up close. This technology will be increasingly used in the gaming sector, but its application in other fields, not just video games, should not be underestimated. Experts in the field of communication or marketing talk about new ways of communicating and selling using this type of technology that will allow users and consumers to experience what will be presented to them. In the industrial sector, at the shop floor level, it is difficult to think, now, about intensive applications for reasons of ergonomics and safety.

The main weaknesses of VR are the high cost of accessories, the still unrealistic quality of VR software, and the disturbance caused to many users while they are using it. Now, there are still many improvements to be made being a developing technology. The first problems to solve are related to motion sickness: when you use virtual reality, you usually find yourself in environments where your virtual avatar moves in a space, while the user is actually still; moreover, the screen is very close to your eyes. All this combined causes a feeling of dizziness, nausea, headaches, and eye fatigue. For all these reasons, it is not advisable to spend long sessions on virtual reality at this moment.

Nowadays, a great emphasis is placed on mixed reality, i.e. identifying devices and software that can provide functionality closer to VR or AR depending on the specific needs of the application context. From a technical point of view, the main challenges we have to face are to (Baratoff and Regenbrecht, 2004):

- Generate high-quality rendering.
- Precisely insert, in terms of position and orientation, virtual objects within a real environment.
- Improve real-time interaction.

All this has meant that there has been a slowdown in sales and that virtual reality is still a niche technology.

4.4 MAIN APPLICATIONS AND SECTORS

Workplace safety appears to be a massive global problem. In line with the recent research trends, Augmented Reality (AR) is one of the emerging technologies involved in the new Industry environments. Presently, it is considered to be ready to be used within the industrial working environments.

During the last two decades, and especially the last few years, augmented reality helps workers manage safety risks on-site and prevent injuries and increase the efficiency of safety.

Augmented reality has the potential to provide far more value to today's workplaces. The possible use cases of AR are different and might be applied in almost all activities taking place in the factories. The main scenarios can be in production, quality control, safety management, maintenance and remote assistance, training of workers, logistics, and design. AR can integrate digital information into how workers perceive the real world, enabling them to seamlessly use that information to guide their choices and actions in real time, as they accomplish tasks.

4.4.1 PERSONNEL TRAINING

Nowadays, the workforce can use Augmented and Virtual (A/V) realities for training, with the result of better interaction between humans and machines. This means that they can:

1) Speed up reconfiguration of production lines.
2) Support shop-floor operators.
3) Implement virtual training for assembling parts.
4) Manage the warehouse efficiently.
5) Support advanced diagnostics integrated into the modules.
6) Interact with the working environment minimising risk (Damiani et al. 2016).

4.4.2   ENGINEERING FIELD

Augmented reality is not just one technology. It is the combination of several technologies that work together to bring digital information into visual perception.

AR services use various device sensors to identify the user surroundings, bringing different experiences to life through portable devices, smart glasses, and AR headsets. That becomes part of daily experience with mobile computing systems.

The digital representation provides both the elements and the dynamics of how an Internet of Things device operates and allows users to view an exact digital representation of a product overlaid on top of the physical one, visualising all kinds of information and details on all the sensors on the device.

Object detection, object tracking, and action recognition are the key issues in finding technical challenges and potential issues when applying computer vision-based approaches in practice.

The use of the Software Development Kit (SDK) is required to ensure the correct operation of AR. An SDK is a set of programmes providing a set of tools, libraries, relevant documentation, code samples, processes, and or guides that allow developers to create software applications on a specific platform (Alam et al., 2014). It allows the programmer to interact more easily with some of the native capabilities on the different OS platforms to create/deliver/serve augmented reality experiences.

It is necessary to state the criteria upon which the different SDKs should be compared to find the SDK that best suits the application. The several major criteria that should be supported by all the features of the wanted application are outlined below.

a) Type of licence. That is what any company should consider first. There are free and commercial/enterprise licences. There is also open-source augmented reality software, to which developers can contribute and add more functions.

b) Supported platforms. Almost any SDK supports both Android and iOS; some AR SDKs are also compatible with the Universal Windows Platform (UWP). There are such augmented reality platforms that allow you to develop AR apps for macOS.

c) Smart glasses support. Unlike smartphones, smart glasses allow hands-free AR experiences. Since smart glasses are becoming more popular (Microsoft HoloLens, Vuzix Blade 3000, Epson Moverio BT-300, and many more), being able to build AR mobile apps compatible with these gadgets is certainly an advantage.

d) Unity support. It is a cross-platform engine that enables you to build apps for multiple platforms including iOS and Android and UWP. Many SDKs are compatible with Unity which brings performance optimisations, tight ongoing synchronisation of features and fixes, and a native Unity workflow that enables developers to create the best AR experiences using a simple authoring workflow and event-driven scripting directly in the Unity Editor.

e) Cloud recognition. In recognising lots of different markers, the development kit must support cloud recognition. Markers are stored in the cloud, while an application does not require much space on a mobile device.

f) On-device (local) recognition. In different situations, there is a possibility of using the on-device markers, without the need to go online to use the app.

g) 3D tracking. Top augmented reality platforms can recognise 3D objects, such as cups, cylinders, boxes, objects, and more. This immensely expands the opportunities for augmented reality in mobile apps.

h) Geolocation. Essential for creating location-based AR apps. One of the examples is if there is a need to add virtual points of interest to the application, an augmented reality platform with geolocation support will be a must.

i) SLAM stands for Simultaneous Localisation and Mapping. SLAM allows applications to map an environment and track their movements in it. One of the examples is the AR mobile app which can remember the position of different things in a workshop or factory and, thus, keeps a virtual object in a certain place while a user moves around the room. Also, this technology can go far beyond adding AR objects. Thanks to SLAM, it is possible to create maps for indoor navigation. GPS does not work indoors, but SLAM does, so this technology has enormous potential.

Following the review of the SDKs, Table 4.1 has been made according to the above criteria. Table 4.1 compares the following AR SDKs: Kudan, Vuforia, EasyAR, Maxst, Wikitude, ARKit, and ARcore.

Exploring SDKs is an important step when moving into AR.

The table shows that all SDKs possess similar features. All of them are very well developed and able to bring top-of-the-class augmented reality experiences. However, the only AR platform currently on the market exclusively focused on the enterprise is Vuforia Studio.

In the working environment, however, the real added value of AR is its ability to make insights from machine intelligence available to the average worker. Most AR systems are currently deployed for tablets – or are smartphone-based – although some are now moving aggressively in the direction of Head-Mounted Displays (HMDs) that allow the worker to access AR data continuously without interrupting a task to refer to a tablet display.

Display Devices. There are three different types of technologies: Video see-through, optical see-through (Albert et al., 2014), and projective-based. The video see-through is closer to Virtual Reality (VR), the digital video of physical world replaces the virtual environment, and the virtual content is superimposed on the video frame. The optical see-through allows us to have a more excellent perception of the real world; the contents in AR are superimposed through mirrors and transparent lenses. The projection-based technology allows projecting the digital content directly on the real object.

Video See-Through. Video see-through is the most economical technique and offers many advantages: (a) the devices that use this technique may be HMD or mobile devices (smartphone or tablet); (b) the current environment is digitised (via video) and it is easier to interact with the real world by superimposing virtual objects; (c) the brightness and contrast of the virtual objects can be easily adapted in the real world; (d) it is possible to match the perception of delay between the real and

TABLE 4.1
The Comparison of Types SDK

Features/Product	Device OS	Development platforms and tools	Cloud image recognition tracking	Location-based services	QR-code	3D tracking	SLAM
Kudan	Android, iOS	Unity3D Plugin, Objective-C API Java API	–	–	+	+	6 DOF camera pose tracking
Vuforia	Android, iOS, UWP	Unity3D NDK Gradle Android SDK Build Tools Android Studio Xcode, Visual Studio	+	+	+ Vumark	+ Model targets	VIO
EasyAR	Android, iOS, UWP	Unity3D, C API, C++11 API, Traditional C++ API Java API Swift API Objective-C API	+ 100K	–	+	3D object tracking	Monocular real-time 6 DOF camera pose tracking
Maxst	Android, iOS, UWP	Unity3D Plugin	–	–	+	You can import map files created with Visual SLAM to enhance desired content	Visual slam- only uses a camera
Wikitude	Android, iOS, UWP	Unity3D Plugin, Cordova plugin, Titanium Module, Xamarin Component	+	+	+	+	6-DoF SLAM
ARKit	Android	Android Studio NDK Ureal Unity3D	–	–	+ Vision framework	3D object detection and persistent experiences	Visual Inertial Odometry VIO, IMU
ARCore	iOS	Xcode	–	–	–	Placing virtual objects with anchors	Concurrent odometry and mapping COM, IMU

the virtual environment. The main disadvantages are: (a) the low resolution of the camera; (b) the limited field of view; (c) in many devices the focus distance cannot be adjusted; (d) In HMD devices, the user may be disoriented because the camera is near the eye positioning.

Optical See-Through. The optical see-through technique is applied to the HMD devices; AR content is mirrored on a curved planar screen. The main advantages are: (a) the display is ideal for a long period of use as it does not create discomfort for the user and leaves real vision unchanged; (b) the user has a direct, unmodified view to the real world, without any delays; the AR objects depend only on the resolution of the display; (c) low energy consumption compared with see-through video. The disadvantages are: (a) the projection of images on the lenses has a contrast and the brightness is reduced; therefore, is not suitable for outdoor use; (b) the reduced field of view can lead to the leakage of the projection from the edges of the lenses; (c) it requires difficult and time-consuming calibration (user- and session-dependent).

Projective. The projective technique is based on the projection of digital content on real-world objects. The advantages are: (a) it does not require lenses to be worn; (b) it allows coverage of large surfaces generating a wide field of vision. The main disadvantages are: (a) the headlamp will be recalibrated if the surrounding environment or the distance from the projection surface changes; (b) it can be only used in indoor environments because of the low brightness and contrast of the projected images (Alam et al., 2017).

4.4.3 OTHER SECTORS

Augmented reality is becoming more and more common within the leading enterprises. The applications of AR and VR technologies on construction (architecture) and industrial design are the following:

Architecture: AV and VR can aid in visualising building projects. Computer-generated images of a structure can be superimposed onto a real-life local view of a property before the physical building is constructed there. Augmented reality can also be employed within an architect's workspace, rendering animated 3D visualisations of their 2D drawings. Architecture sight-seeing can be enhanced with AR applications, allowing users to view a building's exterior to virtually see through its walls, viewing its interior objects and layout.

Industrial Design: AR allows industrial designers to experience a product's design and operation before completion. In general, the augmented and virtual realities are used in the industrial design mostly by the big automobile enterprises as well as the aviation ones.

The areas covered by A/V reality in the industrial systems engineering domain are the following:

1) Product design
2) Logistics
3) Management of production systems
4) Maintenance
5) Safety of production systems

4.4.4   REAL APPLICATIONS

The study (Revetria et al., 2019) presented a methodology for monitoring the stress status of loaded metal shelving to be integrated within an AR environment. The system uses strain gauges to evaluate the stress status of the component, and each operator wearing special virtual reality glasses can visualise the elaborated results. A test to estimate the potentialities of the system, utilising a metal shelf loaded with mono-axial stress as a sample has been carried out, which has shown coherent stress values, which have been correctly displayed by the AR visors.

The advantage of the methodology adopted in this study (Revetria et al., 2019) is the possibility to monitor the stresses in mechanical parts in real time, providing, therefore, benefits in terms of workers' safety and prompting maintenance intervention in case of alert signals.

The study (Damiani et al., 2018) shows that nowadays, virtual training (22%) and maintenance (19%) are the main fields of application of VR and AR technologies.

Many articles analyse the combination of A/V reality and simulation techniques. The study (Streusand et al., 2010) developed an application based on simulation and virtual reality to plan robots' paths. The application also takes into account Computer-Aided Drawing (CAD) and the MATLAB suite. The paper (Hahn et al., 2015) proposes an assistance system based on AR for teaching employees the assembly process of printed circuit boards. The system operates using smart glasses and four characteristic markers, and by looking at QR-codes, the system highlights a component's retrieval location and installation point in the user's field of vision. The study applied in a real company resulted in an errorless performance by each employee (Hahn et al., 2015). Practical support of on-the-job training requires A/V reality technologies.

So, it could be concluded that A/V reality technologies are welcomed in manufacturing, and these solutions have been tested in real industrial settings. Immersive A/V reality technologies open up the possibility of new ways to interact with workers, the shop floor, and the whole enterprise. Workers strengthen their perception of the product and its processes becoming an active part of the manufacturing decision-making as they influence the design of the manufacturing network.

4.5   FUTURE SCENARIOS

Assuming the development of future scenarios, it is necessary to talk about existing problems in all areas of using virtual and augmented reality.

The industrial workshop or factory is quite often a source of potentials risks with automated machines, robots, trucks, chemicals, and threats even when some free-risk activities are engaged. Overconfidence, poor procedure enforcement, and lack of control are the main sources of industrial risk. A/V reality application can be adopted to build a safer environment by promoting a better interaction between people and technology and people in the working environment. These applications should be able to support workers effectively.

The wearable devices used to realise augmented reality have strengths and weaknesses, depending on the purpose of the particular application. For the industrial workplace, glasses are generally handy since they free the operator's hands and are mobile and easily portable. There is a wide variety of Augmented Reality Smart Glasses (ARSG). Some of them have a simple head-up display that serves as a second screen accessible at a glance, whereas others implement more complex solutions such as retinal projection or holographic display.

The main characteristics of AR smart glasses that should be considered when using them are: (1) powering, (2) weight, (3) field of view, (4) battery life, and (5) optics. Let us consider their disadvantages.

Powering. There are two ways of powering ARSG: either through a battery pack or an ordinary computer. For the working environment, battery power is essential, as an operator cannot carry a computer throughout the workday. Existing batteries intended to power the ARSG system are either not energy-intensive enough, or have a lot of weight, which the employee must carry in the appropriate bag for at least several hours.

Weight. This characteristic is essential for the worker themselves since ARSG are meant to be worn more or less the whole day by operators, thus the weight of the ARSG is critical. The weight of a pair of normal glasses is about 20 grams, but no ARSG available today are even close to this weight. The recommended weight can be set at the upper limit of 100 grams (a minimum of five times the weight of normal glasses). Such weight can allow most users to wear the glasses for at least a few hours.

Field of view. The field of view is the area in which virtual objects can be seen via the ARSG and is a crucial parameter as it directly indicates how much information can be shown to the user and where it can be placed. The horizontal field of view is very important. A human's natural field of view is almost 180 degrees horizontally, but today's ARSG are far from matching this. A realistic, acceptable minimum field of view in ARSG is accepted to be 30 degrees (horizontally) (Mladenov et al., 2018).

Battery life. ARSG wear time should be at least eight hours, which corresponds to the industrial operator working day, so it is necessary to have a battery with long service life. For the integrated batteries, the battery life must be at least nine hours, or if they can be fast-charged – four hours (in this case there is a possibility to recharge the battery during the lunch break).

Optics. To visualise information using ARSG, three types of optics can be used – video, optical, and retinal projection. It is considered that for the industrial environment, a product that implements an optical or a retina-based solution should be selected, and video-based solutions should be avoided because video-based solutions have latency in what the user sees compared with what is happening at this moment.

Other essential criteria include:

1) Glasses' camera.
2) Open API.
3) Audio, including microphone and speaker – this is important for communication.
4) Sensors – with types depends on the necessity of the application (for example, gyroscope, compass, GPS, and head or gesture tracking).

5) Controls – should be hands free to allow workers not to be distracted from their immediate job, and usually works through voice commands.

6) Processors – the computer processor should be minimum dual-core.

7) Storage – minimum 30 GB.

8) Memory – minimum 2 GB.

9) Connectivity – Wi-Fi and Bluetooth are necessary to connect the glasses to the Internet or local network, etc.

According to the study (Mladenov et al., 2018) the ARSG that could be used should have the following hardware features: display with a resolution of at least 2 GHz, dual-core CPU, 2 GB RAM, 30 GB internal memory with an external slot, at least a 5 MP camera with 1080p video recording, speaker and noise-cancelling microphone, micro-USB port, Wi-Fi and Bluetooth 4.0 connectivity support. The device should be equipped with buttons, two for scrolling forward and back, one for selection, and one for switching on and off. It should have a voice recognition system using proprietary libraries but is expandable. The operating system should be compatible with the Vuforia SDK. Battery life is to be at least 6 h.

Therefore, we can summarise that the important parameters of ARSG that need further development in the future are:

1) Device cost reduction.

2) The use of most energy-efficient batteries.

3) ARSG weight reduction.

4) Increasing the viewing angle of the ARSG.

5) Lengthening battery life.

6) Improving geolocalisation indoors.

7) Improving device integration (QR connectivity speed, RFID connectivity).

8) The ability for each augmented reality system to work on any software system (Android, iOS, UWP, etc.).

9) The possibility to create a unique set of functions to purchase your own system and to integrate it.

4.6 CONCLUSION

In this chapter, we have considered augmented and virtual reality. From the historical point of view of the emergence of AR and VP, their contribution to the development of Industry 4.0, the field of technology application in the modern industrial sector, as well as future scenarios of the development of AR and VR, their features and shortcomings requiring further development were examined.

REFERENCES

Alam, M. F., Adamidi, E., Hadjiefthymiades, S. 2014. Wireless Personnel Safety System (WPSS), a Baseline towards Advance System Architecture. In *Proceedings of the 18th Panhellenic Conference on Informatics*. 1–6.

Alam, M. F., Katsikas, S., Beltramello, O., Hadjiefthymiades, S. 2017. Augmented and Virtual Reality Based Monitoring and Safety System: A Prototype IoT Platform. *Journal of Network and Computer Applications*, 89: 109–119.

Al-Ahmari, A. M., Abidi, M. H., Ahmad, A., Darmoul, S. 2016. Development of a Virtual Manufacturing Assembly Simulation System. *Advances in Mechanical Engineering*, 8: 1–13.

Albert, A., Hallowell, M. R., Kleiner, B., Chen, A., Golparvar-Fard, M. 2014. Enhancing construction hazard recognition with high-fidelity augmented virtuality. *Journal of Construction Engineering and Management*, 140(7), 04014024.

Azuma, R. T. 1997. A Survey of Augmented Reality. *Presence: Teleoperators and Virtual Environments*, 6: 355–395.

Baratoff, G., Regenbrecht, H. 2004. Developing and Applying AR Technology in Design, Production, Service and Training. In Ong, S. K., Nee, A. Y. C. (eds.) *Virtual and Augmented Reality Applications in Manufacturing*. Springer, London, 207–236.

Damiani, L., Demartini, M., Guizzi, G., Revetria, R., Tonelli, F. 2018. Augmented and Virtual Reality Applications in Industrial Systems: A Qualitative Review Towards the Industry 4.0 Era. *IFAC-PapersOnLine*, 51(11): 624–630.

Damiani, L., Revetria, R., Volpe, A. 2016. Augmented Reality and Simulation Over Distributed Platforms to Support Workers. In *Proceedings - Winter Simulation Conference IEEE Press*. 3214–3215.

Demartini, M., Tonelli, F., Damiani, L., Revetria, R., Cassettari, L. 2017. Digitalization of Manufacturing Execution Systems: The Core Technology for Realizing Future Smart Factories. In *Proceedings of the Summer School Francesco Turco*. 326–333.

Gorecky, D., Khamis, M., Mura, K. 2017. Introduction and Establishment of Virtual Training in the Factory of the Future. *International Journal of Computer Integrated Manufacturing*, 30: 182–190.

Hahn, J., Ludwig, B., Wolff, C. 2015. Augmented Reality-Based Training of the PCB Assembly Process. In *Proceedings of the 14th International Conference on Mobile and Ubiquitous Multimedia. ACM*. 395–399.

Jimeno, A., Puerta, A. 2007. State of the Art of the Virtual Reality Applied to Design and Manufacturing Processes. *International Journal of Advanced Manufacturing Technology*, 33: 866–874.

Kolberg, D., Zühlk, D. 2015. Lean Automation Enabled by Industry 4.0 Technologies. *IFAC-PapersOnLine*, 48–3: 1870–1875.

Mawson, V. J., Hughes, B. R. 2019. The Development of Modelling Tools to Improve Energy Efficiency in Manufacturing Processes and Systems. *Journal of Manufacturing Systems*, 51: 95–105.

Mladenov, B., Damiani, L., Giribone, P., Revetria, R. 2018. A Short Review of the SDKs and Wearable Devices to be Used for AR Application for Industrial Working Environment. In *Proceedings of the World Congress on Engineering and Computer Science 2018 Vol I, WCECS 2018*, October 23–25, 2018, San Francisco, USA. 137–142.

Milgram, P., Takemura, H., Utsumi, A., Kishino, F. 1995. Augmented Reality: A Class of Displays on the Reality-Virtuality Continuum. In *Proceedings of SPIE – The International Society for Optical Engineering*, 2351: 282–292.

Revetria, R., Tonelli, F., Damiani, L., Demartini, M., Bisio, F., Peruzzo, N. 2019. A Real-Time Mechanical Structures Monitoring System Based on Digital Twin, IoT and Augmented Reality. In *Proceedings of the Annual Simulation Symposium. Society for Computer Simulation International*. 18–22.

Streusand, D. B., Steuben, J., Turner, C. J. 2010. Robotic Interfaces Through Virtual Reality Technology. In *ASME 2010 International Mechanical Engineering Congress and Exposition. American Society of Mechanical Engineers*. 419–428.

5 Cloud Computing
Virtualization, Simulation and Cybersecurity – Cloud Manufacturing Issue

Vidosav D. Majstorovic and
Slavenko M. Stojadinovic

CONTENTS

5.1 INTRODUCTION

Cloud computing (CC) has evolved from distributed, parallel, utility and grid computing, and in 2006 the earliest information about it emerged, so that this approach has since developed as a new paradigm. The basis of a new approach (CC) makes up the development of the virtual computer network, with central hardware and software systems, as a platform for providing services to various users. The network that offers a variety of computer resources (hardware/software) to potential users is today referred to as a cloud network (Jianhua et al. 2011; Rakesh and Rinkaj 2019). Also, supercomputers, as a paradigm, are based on the internet and CC, which enable users to dynamically share hardware, software and resource data pool, in conformance with their real needs.

Although more than ten years have passed since the beginning of the development of CC, there still isn't an unambiguous definition of CC; however, the National Institute of Standards and Technology (NIST) has gone the furthest in the development of this discipline, making its definition the most widely used today:

> Cloud computing is a model for enabling ubiquitous, convenient, on-demand network access to a shared pool of configurable computing resources (e.g., networks, servers, storage, applications and services) that can be rapidly provisioned and released with minimal management effort or service provider interaction (Mell and Grance 2011)

The NIST model has five primary characteristics (virtual computing resource, broad network access, on-demand self-service, rapid elasticity, measured service), thus providing a unified integrated cloud service to users. This concept has three service models: Infrastructure-as-a-Service (IaaS), Platform-as-a-Service (PaaS), Software-as-a-Service (SaaS) and four deployment service models (private cloud, community cloud, public cloud and hybrid cloud) (Jianhua et al. 2011; Mell and Grance 2011; Gill 2014); see Figure 5.1.

CC has brought about many benefits for its users but, unfortunately, it has also brought about risks related to the security in this concept application. Today, this is one of the greatest challenges in spreading the application of the concept.

5.2   CLOUD COMPUTING – BASIC CHARACTERISTICS

Cloud computing service models. Cloud computing services have unique characteristics that define the context of possibilities for customers pertaining to his requirements. A survey of such characteristics for SaaS, PaaS and IaaS models is given in Table 5.1 (Liu et al. 2011).

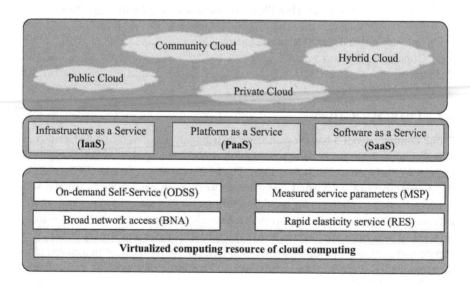

FIGURE 5.1   The NIST model of cloud computing (Mell and Grance 2011).

TABLE 5.1

Cloud Computing Service Models

Service	Model	Capability	Example Services
Software-as-a-Service (SaaS)	Consumer uses vendor's applications, which they manage. Consumer accesses applications over the internet, mainly over the client interface or web browser.	Consumer control is limited by the application configuration, which is the same for all users. Consumer does not have access to the interface for its additional cloud configuration.	Email & Office Document Management, Procurement, Sales, Financial, ERP, Human Resources, Social Networks. *Cloud Manufacturing & I4.0 Platform.*
Platform-as-a-Service (PaaS)	This service allows customers to create and apply their applications, using standard programming languages, as well as tools for platform and infrastructure development.	User has control over deployed application, as well as limited configuration settings for application hosting. However, user does not have access to cloud control and management interfaces.	Application Deployment, Database, Development & Testing, Business Intelligence. *Cloud Manufacturing & I4.0 Platform.*
Infrastructure-as-a-Service (IaaS)	This service allows the user a minimum of computer resources: processors, memories, networks and other users, where the user can apply and start optional software including operating systems and applications.	User manages deployment of application, memory and operating system. Also, network control is possible.	Backup & Recovery, Storage, Platform Hosting, Network. *Cloud Manufacturing & I4.0 Platform.*

Source: Adopted from Liu et al. 2011.

A cloud user can choose one or multiple services from a cloud provider, in conformance with their needs. A specified contract of cooperation between the provider and the user defines the level and type of service, as well as all other parameters of a common cloud model.

Cloud computing deployment models. These models define cloud physical infrastructure context, platforms of use and models applied. Table 5.2 describes deployment models, four in total, such as: public cloud, private cloud, hybrid cloud and community cloud (Liu et al. 2011).

The customer has three cloud service models available (SaaS, PaaS and IaaS), which are deployed in the cloud, and the cloud can be: private, community, public and hybrid. We are particularly interested in their relationships with business systems: private cloud is a model most suitable for large organizations/companies, whereas community cloud suits small and midsize organizations best.

TABLE 5.2

Cloud Computing Deployment Models

Deployment Model	Description
Private cloud	Provider provides the user with exclusive access and rights to the use of infrastructure and computer resources. This implies that cloud infrastructure and computer resources are rented out as an outsource to a third-party hosting organization.
Public cloud	In this model the provider provides cloud services to a set of diverse cloud service users. They are provided over the public network which may be wireless or wireline.
Community cloud	In this model the cloud service is provided to a group of users who have a common interest: goals, security, privacy and compliance policy. Group cloud can be implemented at the location of a single user, or at the provider as a hosting organization.
Hybrid cloud	This model is composed of two or more clouds and is a combination of above-mentioned models, on or off-site. In this model, each cloud is a separate entity.

Source: Data from Liu et al. 2011.

Starting from steady development of the IT environment, it can be said that the enterprise business model will change too, in conformance with deployment models.

Cloud computing reference architecture and taxonomy. Reference architecture and taxonomy are also very important characteristics for cloud providers (CP). NIST recommended reference architecture for cloud computing (CC); see Figure 5.2 (Mell and Grance 2011; Liu et al. 2011). Cloud computing architecture has five primary actors that are used to define their functional activities. Also, various deployment models of computer resources were developed by which the cloud providers provide their services.

A taxonomy model for cloud computing services, which describes a cloud model, which is based on the NIST reference architecture is given in Figure 5.3 (Mell and Grance 2011; Liu et al. 2011).

Taxonomy for the cloud functional model that determines various roles in the cloud computing IT environment, activities associated with these roles, components (concrete processes, activities or tasks to perform so as to achieve the defined goals) and a sub-component (connected component module) are shown in Table 5.3 (Liu et al. 2011).

Virtualization technologies. Virtualization in cloud computing helps to start the same application multiple times simultaneously, by one or multiple cloud clients, creating a virtual model. This process ensures steady state in the case of multiple users starting the same application in a secure environment. Virtual machines (VMs) and virtual machine monitors (VMMs) are components of virtualization that give flexible and adjustable computer resources (Grobauer et al. 2011). VMMs allow virtual hardware resource allocation such as memory, CPU, hard disk and virtual network interfaces for each VM. Also, VM migration is one of the greatest advantages of using virtualization in cloud computing, which enables transferring of a VM port from one hardware server to another in order to finish or complete the scheduled process. The more virtualization cloud-based platforms that operate in a multi-tenant cloud environment, the more successful the adoption of such a green computing

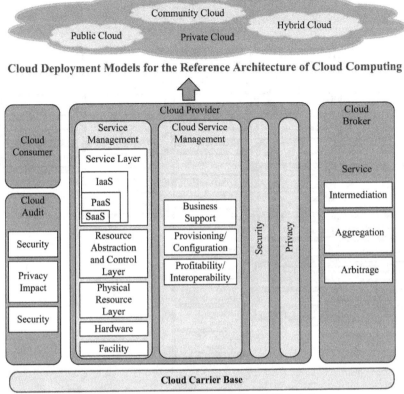

Cloud Deployment Models for the Reference Architecture of Cloud Computing

Cloud Reference Architecture for Cloud Computing

FIGURE 5.2 Cloud computing reference architecture with different deployment models. Adapted according to (Mell and Grance 2011; Liu et al. 2011).

architecture strongly depends on its security assurance mechanisms; see Figure 5.4 (Grobauer et al. 2011).

Web services. Web services and service-oriented architecture (SOA), of the cloud system, is a model most widely used today (Rakesh and Rinkaj 2019). SOA provides a framework that enables communication between the systems through the interaction services they are engaged in. Hypertext Transfer Protocol (HTTP), Representational State Transfer (REST), eXtensible Markup Language (XML), Simple Object Access Protocol (SOAP), on-demand self-service, Universal Description, Discovery and Integration (UDDI), Web Services Description Language (WSDL) and Universal Description, are some of the most widely used technologies for implementing the web services (Shawish and Salama 2014).

Internet-based technology. Currently, internet-based technology represents a basis for the cloud development and application, based on broadband networks, enabling rapid access to cloud services. The most popular internet-based technology is the World Wide Web (WWW), through which IT resources are connected, providing best solutions to users (Fernandes et al. 2014).

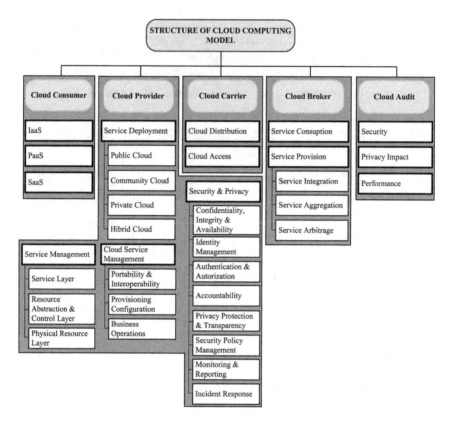

FIGURE 5.3 Cloud computing taxonomy. Adapted according to (Mell and Grance 2011; Liu et al. 2011).

Multi-tenant technology. This technology enables multiple users to access the same cloud application simultaneously. In the process each tenant manages the application functions according to their own needs (Fernandes et al. 2014). Also, this technology provides system scalability and metered usage for each user.

Data-centre technology. Working of cloud computing is supported by data-centres with mass memory servers expressed in Gigabit Ethernet. On the other hand, these servers enable virtualization in cloud computing, thereby opening up abundant possibilities for users such as: unlimited hardware, rapid broadband access, services acceptable or low costs, all defined by legislation, which is of utmost importance for business users (Gajbhiye and Shrivastva 2014).

5.3   CLOUD COMPUTING AND SECURITY MODELLING

The issue of security is especially important in the cloud computing model, because the environment these models are running in is exposed to various threats. For this reason, special models of security requirements related to the components and overall cloud computing model are developed (Rakesh and Rinkaj 2019).

TABLE 5.3

Cloud Computing Active Roles and Their Activities

Roles of CC	Description	Activities
Cloud user	A person or organization that utilizes and pays for services provisioned by the provider. They define common business relationship by the mutual contract.	A cloud service is delivered to the cloud user according to one of the models – SaaS, PaaS, IaaS.
Cloud provider	Cloud provider is a person or organization responsible for providing services to cloud users. Cloud provider is the owner of computer infrastructure and software and also defines the framework for providing services primarily via the internet.	Cloud provider also provides infrastructure and services management in the informatics cloud – service deployment, service termination, cloud service management as well as privacy and security.
Cloud carrier	Cloud carrier provides connection with cloud users, via service transfer from cloud provider. Depending on the service level offered by the provider to the user, an appropriate cloud carrier model is used.	For providing connection and transfer of cloud services between the provider and the user.
Cloud broker	Cloud broker that negotiates business relationships between cloud provider and their cloud users. Cloud broker offers adapted service packages, its feasibility from different providers, so that the cloud user can obtain better and larger service. Cloud broker manages delivery and its performances like the cloud carrier.	VAT increase for cloud partners – providers and users.
Cloud auditor	Cloud auditor is an independent organization performing neutral assessment of the cloud model application, against the services, performances, operation, privacy and security. Cloud auditor performs assessment based on objective evidence, against the defined standards and regulatory requirements.	Assessment is made in compliance with standards and regulatory requirements of applied cloud service model.

Source: Data from Liu et al. 2011.

5.3.1 CLOUD COMPUTING LAYERED MODEL

One of the widely used representations of the cloud architecture components is the multi-layered stack model that allows depicting technology associated with each layer representing as-a-service delivery model; see Figure 5.5 (Mogull et al. 2017). The basic layer structure includes three parts: application and interface layer, platform layer and infrastructure layer.

Application and interface layer. These two layers, first of all, support the SaaS model to use various web services such as: HTML, SOAP, WSDL, JavaScript, AJAX, REST, CSS, etc. (Rakesh and Rinkaj 2019). The user accesses them by web

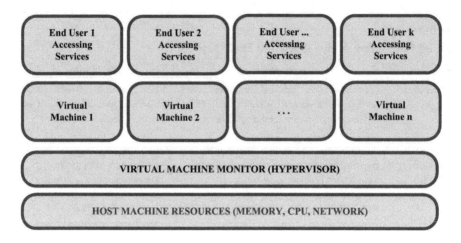

FIGURE 5.4   Virtualization framework for cloud.

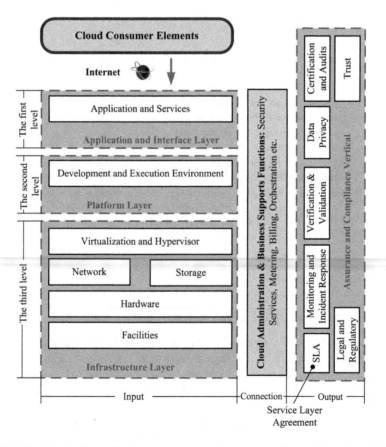

FIGURE 5.5   Model of layer for cloud computing. Adapted according to (Mogull et al. 2017).

browsers using all elements of security along with the protocol such as public-key cryptography, OAuth and OpenID.

Platform layer. Software development, its testing and application through cloud computing is done according to user requirements, in this layer. Additionally, the platform layer provides the needed operating system to its users: IDE, .NET, Java SDK, Google App Engine (Rakesh and Rinkaj 2019).

Infrastructure layer. Optimum computer resource utilization in cloud computing is achieved by the development and application of virtual models, running in the abstraction layer using multiplexing (Rakesh and Rinkaj 2019). VMM or Hypervisor manages virtualized resources such as processors, memory and I/O devices.

5.3.2 Security Taxonomy and Requirements for Cloud Computing

Taxonomy defines a model of creating security requirements for cloud computing, which are based on 12 security threats that may emerge in cloud computing (Mogull et al. 2017; Cloud Security Alliance 2016). The security threats are underlying the proposed countermeasures that increase the security of the overall CC system and reduce its vulnerabilities.

NIST has defined confidentiality, integrity and availability as primary security requirements in cloud computing, which also holds for any other information security management system (Liu et al. 2011). This way, additional identity verification, its authorization, responsibility and privacy are obtained as security requirements in CC (Mogull et al. 2017).

5.3.3 Cloud Security Threats and Vulnerabilities

The NIST Glossary of Security Terms defines threat as,

> Any circumstance or event with the potential to adversely impact organizational operations (including mission, functions, image, or reputation), organizational assets, individuals, other organizations, or the Nation through an information system via unauthorized access, destruction, disclosure, modification of information, and/or denial of service

(Kissel 2013) Cloud Security Alliance 2019; Grobauer 2013; Hashizume et al. 2013 provide a summary of research results related to identification and analysis of threats for CC, representing a framework for future research. Also, countermeasures to prevent threats in CC are included.

Furthermore, the NIST Glossary of Security Terms defines vulnerabilities as, "Weakness in an information system, system security procedures, internal controls, or implementation that could be exploited or triggered by a threat source" (Kissel 2013).

5.4 CLOUD-BASED MANUFACTURING

Cloud manufacturing (CM) represents integration of cloud and service computing with manufacturing processes. This new model of technological systems is experiencing intensive development today, in terms of both research and application.

Cyber-Physical Systems (CPS) are the next paradigm that integrates computation, networking and physical processes, which are directly monitored and supervised over the internet and applied in manufacturing. This necessitates integration of cloud manufacturing and CPS for the sake of manufacturing services realization. This also means that now we are directly monitoring and operating machine tools in manufacturing cloud. As a result, we have a new paradigm of Cyber-Physical Manufacturing Clouds (CPMC) (Liu et al. 2017). A CPMC is a type of manufacturing cloud where machining tools can be directly monitored and operated over the internet from the clouds. Many existing manufacturing clouds provide manufacturing services but do not allow their machining tools to be monitored and operated directly from the clouds.

Starting from the NIST definition (Mell and Grance 2011), many researchers complemented the definition of CM, so that it now reads:

> Cloud Manufacturing (CM) is a customer-centric manufacturing model that exploits on-demand access to a shared collection of diversified and distributed manufacturing resources to form temporary, reconfigurable production lines which enhance efficiency, reduce product lifecycle costs, and allow for optimal resource loading in response to variable-demand customer generated tasking (Wu et al. 2013a)

It can be concluded that cloud computing is based on the interaction of three stakeholders: consumers, application providers and physical resource providers.

Users are the consumers in CM – these individuals or groups have the need to manufacture something, but do not possess the capabilities to do so, or they possess the capabilities but stand to gain a competitive advantage by utilizing CM. The cloud-based application layer is responsible for managing all aspects of the CM environment and interprets user requirements into data required for production of the desired objects. Physical resource providers (PRPs) own and operate manufacturing equipment, including but not limited to machining technologies, finishing technologies, inspection technologies, packaging technologies and testing resources (Wu et al. 2013b).

For example, the ManuCloud architecture provides users with the ability to utilize the manufacturing capabilities of configurable, virtualized production networks, based on cloud-enabled, federated factories, supported by a set of Software-as-a-Service applications (Meier et al. 2010).

On the other hand, the ManuCloud connects customers with production sites and customized products via inter- and intra-factory connections.

For the mentioned CM model in (Xu 2012) a four-layer CM framework is shown, involving: manufacturing resource layer, virtual service layer, global service layer and an application layer; see Figure 5.6.

Thus, for example, the Manufacturing Resource Layer (Xu 2012) involves the physical manufacturing resources and capabilities of the shop floor (machine tools, robots, AGVs, CMM), which are ultimately provided to the customer in Software-as-a-Service (SaaS) and Infrastructure-as-a-Service (IaaS) delivery models. The Virtual Service Layer identifies, virtualizes and packages the resources as CM services (models of design, planning and inspection of products and processes). Service Layer can run in full or partial operational mode, as defined by the user.

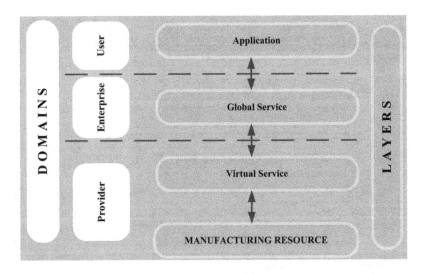

FIGURE 5.6   Layers by domain (Xu 2012).

The most complex layer structure shown in Figure 5.7 has seven layers (Zhang et al. 2011).

In the first layer, there are elements of the enterprise structure relating to the customers, vendors or the organization viewed as a part of the supply chain. For the most part, these are business activities, whereas manufacturing elements refer mainly to drawings as an input element for technical documentation. The second and third layer represent CM framework for production, and in the next layer this is defined for a concrete organization. The central and most important layer is the one that connects the elements of workshop activities according to the CC model. These are all elements of production and technological resources scheduling and management, including quality maintenance management. The next layer performs the virtualization of all manufacturing resources from digital twins. The sixth layer monitors the process elements in real manufacturing, and in the final layer the analysis and synthesis of collected information is done for supervisory control and management of the overall CM.

The text below presents three chosen examples of CM that explain in a specific way the above-mentioned CC elements applied in CM.

5.4.1 CASE STUDY 1: CLOUD-BASED DESIGN AND ENGINEERING ANALYSIS OF SOFTWARE TOOLS (WU ET AL. 2016)

The first example involves engineering design, which is the basis for creating a CM model, and proceeds through several stages: conceptual design, embodiment design and detail design. The starting parameters for conceptual design are defined by identifying customer and market demands and needs. Based on these elements, product characteristics and its concept are defined. In this stage, the product lifetime can be defined too. The activities involve usage of various techniques and product planning tools such as questionnaires and customer surveys and market research.

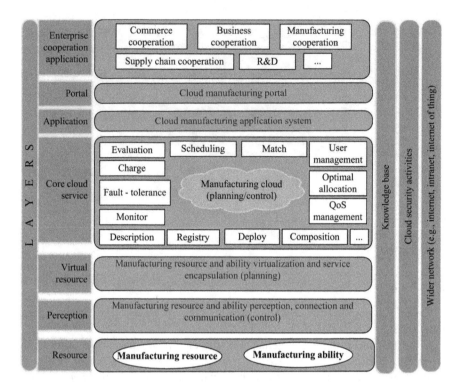

FIGURE 5.7   The most complex layer structure. Adapted according to (Zhang et al. 2011).

If we commence from the CC concept, for this stage – conceptual design – the Salesforce Market Cloud software can be used. It collects data from the cloud in a digital marketing model and product data identified by potential users as defined by their questionnaires and surveys. On the other hand, it cannot collect data obtained from social networks such as Google, Twitter and others.

In the next step, the choice is made between products defined in the previous stage. The procedure employs various techniques and tools such as: modelling, brainstorming, reasoning, representation, morphological matrices, Pugh's method and other aids of decision-making. Here, it is important to consider all aspects of costs (manufacturing, exploitation), performances and product reliability. For this stage there are several CC technology-based tools, but the technique for automatic product selection from the previous stage is still lacking.

Finally, in the last stage – embodiment design and detail design – the final structure and geometry of the product are defined. For several decades, this area has been using a variety of digital tools referred to as computer-aided design systems. Today, for CC support this field utilizes Autodesk 360 and AutoCAD 360, which have the following features: (a) share documents using cloud storage services such as Dropbox, Box and others; and (b) view, create and edit 2D drawings in the DWF and DWG formats on multiple devices using a web browser, smartphone, tablet and desktop computer. Among the cloud-based 3D CAD tools is Fusion 360, which includes the following unique features through the cloud: freeform shape modelling, solid modelling, parametric modelling and mesh modelling. One of the best-known tools for this area is Teamcenter, a cloud-based tool design developed by Siemens (PLM software). The features of which include: (a) enterprise-grade ICT infrastructure and resources; (b) fast deployment options and simplified scalability for lower cost; and (c) use of the IaaS platform. Other examples include: Dassault Systems has developed cloud-based solutions such as cloud-based Solidworks, CATIA and Simulia via the 3D Experience Platform. Specifically, the features of Solidworks on the cloud include CAD animation, revision control, part and assembly modelling, tolerance analysis, CAD library and interference check.

5.4.2 CASE STUDY 2: SCHEDULING BY CM MODEL (HU ET AL. 2019)

The next example is related to the most important, initial element of the manufacturing scheduling and management – manufacturer scheduling – where there is the rational allocation of manufacturer resources under the cloud manufacturing platform, and it can make users in the platform get the normal, good manufacturing services. The scheduling process of manufacturers under the cloud manufacturing environment model is given in Figure 5.8 (Hu et al. 2019).

Termination is a complex model, which reflects on its modelling for CC, so that providers are making efforts to meet increasingly complex user demands that are becoming diversified. They are additionally complicated when the customer

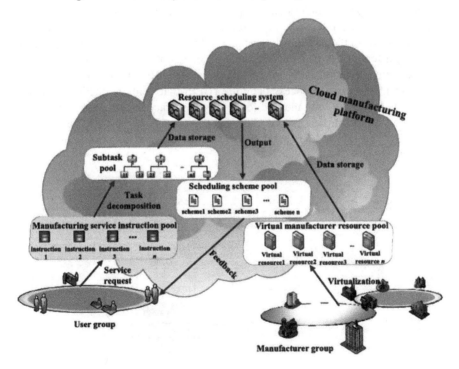

FIGURE 5.8 Scheduling in CM concept. Reprinted with permission from Elsevier (Hu et al. 2019).

demands are included in this model. The major goal of all models is to optimize, or more precisely, minimize product or service delivery time, which refers to manufacturing. This way, all organizational business parameters are increasing.

Here, we give an example of a complex model of termination that contains five categories of manufacturer scheduling indexes, including manufacturer task load (FZ), manufacturing efficiency (XL), manufacturing resource richness (MR), IoT match degree (LoT) and task reliability (RK). This means that the user can utilize one or more criteria for the terminating process in conformance with their own production model. It is especially important that re-scheduling can also be done in the case of urgent jobs. Increasing the reliability of scheduled and executed jobs is the basic parameter of the success of this model.

5.4.3   Case Study 3: Cyber-Physical Manufacturing Metrology Model (CP3M)

The third example is of current research linked to CM, a part related to cloud manufacturing metrology (Majstorović et al. 2017). The research considers the idea of developing the area of manufacturing metrology as a constituent part of CM with the support of a coordinate metrology machine (CMM), which has already become the Cyber-Physical Model. The structure of our research model – CP3M – is presented in Figure 5.9, and includes two basic components: physical and virtual.

The entire model embraces all elements of scheduling, preparation and realization of inspection for box-like parts on CMM, using the developed elements of the structure from the previous figure: (a) module for definition and recognition of geometrical features (GF) from CAD/GD&T model of the measuring part, where we used them for definition of metrological feature (MF); (b) module for building of intelligent inspection process planning (IIPP) that contains methods for prismatic parts presented, and method for freeform surfaces application; (c) CMM – generation of control data list for CMM that is transferred to CMM using cloud technology; and (d) module for analysis of the results and generation of the final measuring reports.

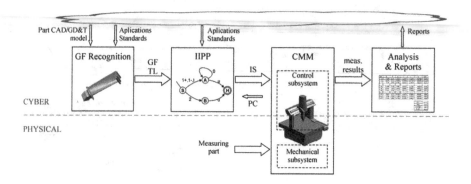

FIGURE 5.9 Cyber-Physical Manufacturing Metrology Model (CP3M) Framework (Majstorović et al. 2017).

The main elements of IIPP are ontology knowledge base, optimization model based on ant's colony optimization, visualization, verification and simulation. The developed model of engineering ontology uses for linking standardizes the form of tolerance and geometry of prismatic parts, i.e., geometrical features. The proposed ontology model is implemented in the Protégé software in the example of prismatic measuring parts (Figure 5.10). A part of other main IIPP elements is shown in Figure 5.11.

The application of ant's colony optimization (ACO) in a coordinate metrology is based on the solution of the travelling salesman problem (TSP), where the set of cities that the salesman should pass through within the shortest possible path corresponds to the set of points of a minimal measuring path length expressed as a point-to-point path.

The purpose of inspection simulation is a visual check of the measurement path from the standpoint of the collision avoidance and generating measurement protocol, as well as the control data list for the process of the measurement execution. Simulation of the path is performed in the Pro/Engineer software version Wildfire 4.0 (PTC Creo) and MatLab software. For the simulation of the path in the software, the Manufacturing module was used within the CMM submodule as well. Figure 5.12a shows the path of inspecting the parameters of the plane, as well as the path of inspecting the parameters of the cone (Figure 5.12b).

Cloud services within the organization provide the necessary information for integration of knowledge and data from various phases in product design and manufacturing/metrology into inspection planning and make available information about inspection results to all interested parties in the product lifecycle.

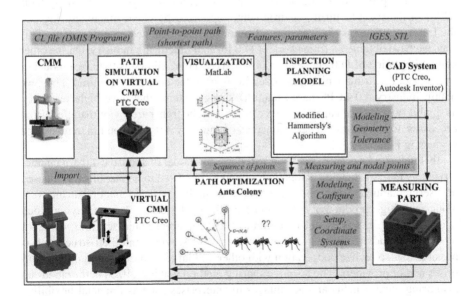

FIGURE 5.10 Implemented taxonomy by engineering ontology in Protégé software (Stojadinovic and Majstorovic 2019).

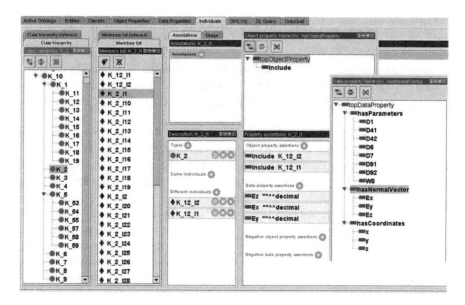

FIGURE 5.11   A part of IIPPP for optimization, visualization, simulation and verification. Adapted according to (Stojadinovic, Zivanovic and Slavkovic 2019).

5.5   INDUSTRY 4.0 AND CLOUD MANUFACTURING

New models of information technologies, applied in new concepts of technological systems, provided the definition of a new paradigm: Cyber-Physical Systems (CPS) in the manufacturing environment, supported by cloud manufacturing and the Internet of Things (IoT) as a new framework for advanced manufacturing. This way, we have obtained a framework of I4.0 for manufacturing industry. On the other hand, integration of I4.0 elements (43 in total) and CM allows us to have a model of smart manufacturing, which is realized in smart value chains in real time. Thus, we obtain Industry 4.0's CPS-based manufacturing systems that have high flexibility, adaptability, real-time capability and can achieve the transparency of manufacturing processes. Comparison between concepts of Industry 4.0 and cloud manufacturing is given in Figure 5.13.

Technological systems operating based on the model of Industry 4.0 are characterized by manufacturing personalized products that used to be manufactured based on a serial manufacturing model, have superior quality, reduced manufacturing costs along with high productivity (Liu and Hu 2017). The basic idea underlying cloud manufacturing in this model is to connect and integrate manufacturing resources of various cloud enterprises, so that in the form of services and their structure it is possible to also realize cooperation with other and large resources from other enterprises and/or providers. The resource sharing method that is realized in the cloud can also bring huge benefits and advantages to enterprises such as financial flexibility, business agility and direct access to innovations (Yu et al. 2015). This implies that Industry 4.0 represents a highly digitalized and networked manufacturing platform

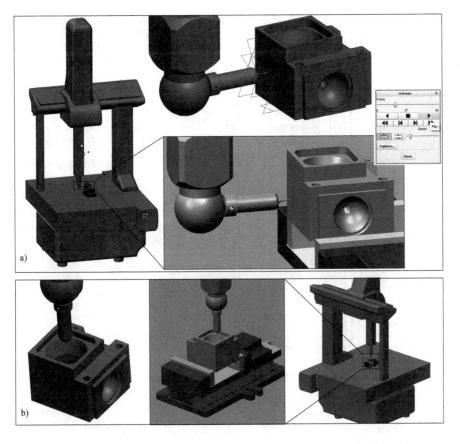

FIGURE 5.12 Measurement path simulation on virtual CMM in CAD/CAM environment (Stojadinovic, Zivanovic and Slavkovic 2019).

that meets the requirements for cooperation between enterprises and business partners in an appropriate and agile manner (Yu et al. 2015).

So, it can be concluded that there are two different platforms that integrate I4.0 and CM, which differ in the following characteristics: goal, operational mode, operational technologies and platform architecture. The CM platforms are seeking to enable a full sharing of all resources, especially manufacturing ones, and at the same time to be open for other enterprises to join. Also, it is the goal of the CC platform to ensure complete support for industrial business processes and compliance business networks, smart factories and the lifecycle of a product (Liu and Hu 2017).

The next characteristic is the operational mode which is coupled with the organizational business model for CM. On the one hand, CC in I4.0 is an open platform, so that basically we have two different types of platforms regarding business model (Industry 4.0 Working Group 2013; Li et al. 2010) which is, as a rule, dynamic. This leads to the need to solve different issues: regulatory requirements, intellectual property and know-how, monitoring of business operations, safety and security.

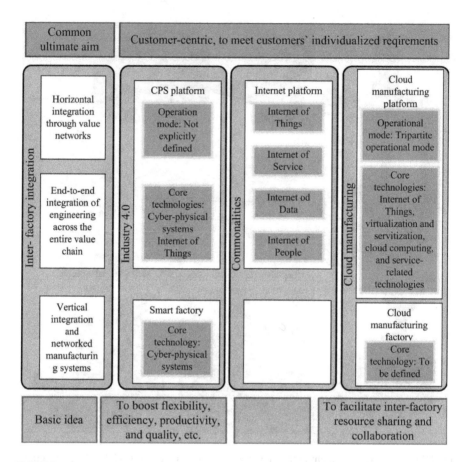

FIGURE 5.13   Connection between Industry 4.0 and CM (Liu and Hu 2017).

Both platforms support two common core technologies: Cyber-Physical Systems and IoT. On the other hand, depending on business goals, the CM platform can include other technologies too: virtualization and servitization, service-related technologies, etc.

Finally, the last characteristic is architecture. CC architecture has not been defined yet, unlike CM where it is adopted that it is layer-based. Such an approach for CM allows us to develop and apply the CM model for various manufacturing organizations (Liu and Hu 2017).

5.6   CONCLUDING REMARKS

Industry 4.0 as a model has developed from the concept of digital manufacturing and the methods to support it. It has been developing for several decades and now with the help of the internet and other new technologies its networking is performed at a higher digitalization level. However, at the same time, manufacturing organizations are increasingly connecting with a cloud, making their products individual customer-oriented, along with the optimization of all resources, including costs. This is the I4.0 model in practice. It can be then concluded that CM is moving towards evolution of transition to a full I4.0 model (43 elements).

It is to be expected that CM, based on the full I4.0 concept, will be "manufacturing-as-a-service". This implies that in the CM operating model the manufacturing process based on "manufacturing" cooperation between smart factories worldwide operating in the cloud would be realized.

REFERENCES

Cloud Security Alliance. 2016. The treacherous 12 - cloud computing top threats in 2016. Tech. Rep. https://downloads.cloudsecurityalliance.org/assets/research/top-threats/Treacherous-12_Cloud-Computing_Top-Threats.pdf (Accessed on June 24, 2019).

Cloud Security Alliance. 2019. Top threats to cloud computing. Tech. Rep. V1.0. Cloud Security Alliance. https://cloudsecurityalliance.org/topthreats/csathreats.v1.0.pdf (Accessed on June 24, 2019).

Fernandes, D., et al. 2014. Security issues in cloud environments: A survey. *Int. J. Inf. Security* 13:113–170.

Gajbhiye, A. and K. Shrivastva. 2014. Cloud computing: Need, enabling technology, architecture, advantages and challenges. In *Proceedings of 5th International Conference - Confluence the Next Generation Information Technology Summit (Confluence)*, 1–7. Institute of Electrical and Electronics Engineers (IEEE).

Gill, K. 2014. The history of cloud computing and cloud storage. www.linkedin.com/pulse/20140602173917-185626188-the-history-of-cloud-computing-and-cloud-storage (Accessed on June 23, 2019).

Grobauer, B., et al. 2011. Understanding cloud computing vulnerabilities. *IEEE Secur. Priv.* 9:50–57.

Grobauer, B., et al. 2013. Understanding cloud computing vulnerabilities. *IEEE Secur. Priv.* 9:50–57.

Hashizume, K., et al. 2013. An analysis of security issues for cloud computing. *J. Internet Serv. Appl.* 4:1–13.

Hu, Y., et al. 2019. Scheduling of manufacturers based on chaos optimization algorithm in cloud manufacturing. *Rob. Comput. Integr. Manuf.* 58:13–20.

Industrie 4.0 Working Group. 2013. Securing the future of German manufacturing industry—recommendations for implementing the strategic initiative. acatech, Munich, Germany. www.acatech.de/fileadmin/user_upload/Baumstruktur_nach_Website /Acatech/root/de/Material_fuer_Sonderseiten/Industrie_4.0/Final_report__Industrie_4.0_accessible.pdf (Accessed on June 25, 2019).

Jianhua, C., et al. 2011. Study on the security models and strategies of cloud computing. *Procedia Eng.* 23:586–593.

Kissel, R. 2013. Glossary of Key information security terms. In NISTIR 7298 revision 2. National Institute of Standards and Technology. http://nvlpubs.nist.gov/nistpubs/ir/2013/NIST.IR.7298r2.pdf (Accessed on June 24, 2019).

Liu, F. et al. 2011. *NIST Cloud Computing Reference Architecture (SP 500-292)*. National Institute of Standards & Technology, Gaithersburg, MD 20899-8930. USA. http://ws6 80.nist.gov/publication/get_pdf.cfm?pub_ id=909505 (Accessed on June 24, 2019).

Liu, X., et al. 2017. Cyber-physical manufacturing cloud: Architecture, virtualization, communication, and testbed. *J. Manuf. Syst.* 43:352–364.

Liu, Y. and H. Hu. 2017. Industry 4.0 and cloud manufacturing – a comparative study. *J. Manuf. Sci. Eng, Trans ASME* 139:1–8.

Li, B., et al. 2010. Cloud manufacturing: A new service-oriented manufacturing model. *Comput. Integr. Manuf. Syst.* 16:1–8.

Majstorović, V., et al. 2017. Cyber-physical manufacturing metrology model (CPM3) for sculptured surfaces – turbine blade application. *Procedia CIRP* 63:658–663.

Meier, M., et al. 2010. ManuCloud: The next-generation manufacturing as a service environment. *Eur. Res. Consortium Inf. Math. News* 83:33–34.

Mell, P. and T. Grance. 2011. The NIST definition of cloud computing (SP 800-145). Tech. Rep., National Institute of Standards & Technology, Gaithersburg, MD. http://nvlpubs. nist.gov/nistpubs/Legacy/SP/nistspecialpublication800-145.pdf (Accessed on June 23, 2019).

Mogull, R., et al. 2017. *Security Guidance for Critical Areas of Focus in Cloud Computing.* Cloud Security Alliance. https://downloads.cloudsecurityalliance.org/assets/research/ security-guidance/security-guidance-v4-FINAL.pdf (Accessed on June 24, 2019).

Rakesh, K. and G. Rinkaj. 2019. On cloud security requirements, threats, vulnerabilities and countermeasures: A survey. *Computer Science Review* 33:1–48.

Shawish, A. and M. Salama. 2014. Cloud computing: Paradigms and technologies. In *Inter-Cooperative Collective Intelligence: Techniques and Applications*, 39–67. Ed. F. Xhafa and N. Bessis. Springer-Verlag Berlin Heidelberg 2014.

Stojadinovic, M. S. and V. D. Majstorovic. 2019. *An Intelligent System for CMM Inspection Planning of Prismatic Parts.* Springer International Publishing, Cham, Switzerland.

Stojadinovic, M. S., T. S. Zivanovic and N. R. Slavkovic. 2019. Verification of the CMM measuring path based on the modified Hammersly's algorithm. In *Proceedings of the 12th International Conference on Measurement and Quality Control - Cyber Physical Issue*, 25–38. Ed. V. D. Majstorovic and N. M. Durakbasa. Springer International Publishing, Cham, Switzerland.

Wu, D., et al. 2013a. Cloud manufacturing: Drivers, current status, and future trends. In *Proceedings of the ASME 2013 International Manufacturing Science and Engineering Conference (MSEC13)*, Paper Number: MSEC2013-1106. American Society of Mechanical Engineers.

Wu, D., et al. 2013b. Cloud manufacturing: Strategic vision and state-of-the-art. *J. Manuf. Syst.* 32:564–579.

Wu, D., et al. 2016. A survey of cloud-based design and engineering analysis software tools. In *Proceedings of the ASME 2016, International Design Engineering Technical Conference & Computers and Information in Engineering Conference - IDETC/CIE*, 1–9. ASME.

Xu, X. 2012. From cloud computing to cloud manufacturing. *Rob. Comp. Integr. Manuf. Syst.* 28:75–86.

Yu, C., et al. 2015. Computer-integrated manufacturing, cyber physical systems and cloud manufacturing-concepts and relationships. *Manuf. Lett.* 6:5–9.

Zhang, T., et al. 2011. Cloud manufacturing: A computing and service-oriented manufacturing model. *Proc. Insti. Mecha. Engi, B: J. Eng. Manuf.* 225:1969–1976.

6 Industry 4.0 and Lean Supply Chain Management

Impact on Responsiveness

Miguel Núñez-Merino, Juan Manuel
Maqueira-Marín, José Moyano-Fuentes,
and Pedro José Martínez-Jurado

CONTENTS

6.1   INTRODUCTION

In recent years, Supply Chain Management (SCM) has come to be considered a key factor in increasing organizations' effectiveness, efficiency, and competitiveness (Frohlich and Westbrook 2001; Gunasekaran, Lai and Cheng 2008).

In this line, great interest has arisen in both the academic and professional worlds around the application of the Lean Management (LM) principles to the Supply Chain (SC) (Wee and Wum 2009; Marodin et al. 2017; Tortorella, Miorando and Marodin 2017). LM originally grew out of the production perspective (Womack, Jones and Roos 1990; Womack and Jones 1996) but has evolved into a much wider ranging management system and spread to other company areas and the SC (Lamming 1996). LM seeks the systematic reduction of the enormous amount of wastage that occurs in the majority of organizations and has demonstrated that reducing the activities that do not contribute any value to an organization and throughout the SC can enhance business results and the ability to gain a competitive advantage (Hines, Holweg and Rich 2004; Shah and Ward 2007). Applying LM to the SC to optimize all the activities from the customer's point-of-view is known as Lean Supply Chain Management (LSCM), and consists of directly linking organizations' upstream and downstream flows of goods, services, finances, and information so as to reduce cost and waste by efficiently and effectively pulling what is required to meet the needs of individual customers (Womack and Jones 1996; Martínez-Jurado and Moyano-Fuentes 2014; Swenseth and Olson 2016; Moyano-Fuentes, Bruque-Cámara and Maqueira-Marín 2019).

Furthermore, in the current competitive environment, and especially since the emergence of Industry 4.0 (I4.0), organizations must be able to gather, analyze, synthesize, interpret, and coordinate a large volume of data along the entire value flow on both the internal and SC levels. In the context of the SC, the concept of I4.0 can be defined as the set of manufacturing technologies implemented in an SC to address the trends of digitization, autonomization, transparency, mobility, modularization, network-collaboration, and interaction between products, processes, suppliers, and customers (e.g., Pfohl, Yahsi and Kurnaz 2017; Buer, Strandhagen and Chan 2018). I4.0 is, therefore, driving an enormous change in SCM by bringing improvements to visibility and the speed with which information is being made available to SC members, among other things (Handfield and Linton 2017).

The goal of achieving efficiency and customer satisfaction through LSCM can, therefore, be more easily achieved with Information Technologies (IT) (Qrunfleh, Tarafdar and Ragu-Nathan 2012) and, more specifically, the IT used in I4.0 (Sanders, Elangeswaran and Wulfsberg 2016). In fact, when some specific IT are intensively applied in productive environments by digitizing products and processes (Kang et al. 2016; Monostori et al. 2016), there is an impact on the responsiveness of companies and the SCs in which they are involved, with a noticeable improvement in their competitiveness (Tortorella and Fettermann 2018). SC responsiveness can be characterized by flexibility and the speed with which companies respond to environmental changes (Schmidt et al. 2015; Satoglu et al. 2018). I4.0 includes a wide range of technologies in this line that can contribute to companies' production systems and make SCs increasingly more autonomous, precise, flexible, and dynamic (Tortorella and Fettermann 2018). Moreover,

IT are tools that firms use in conjunction with other resources and capabilities to gain a competitive advantage (Powell and Dent-Micallef 1997; Carr 2004).

For all these reasons, LSCM and I4.0 could be complementary approaches as their joint application could enable them to impact the SC's responsiveness by improving both its flexibility and its response speed. The objective of this chapter is, precisely, to provide an overview of the key aspects and implications of the relations between I4.0 and LSCM in the area of improvements to SC responsiveness, which is one of the key outcomes of SCM at the global level. The relevance of this article, therefore, revolves around LSCM environments, which are characterized by their poor mass-customization capability and limited adaptation to changing environmental requirements (Vonderembse et al. 2006) and the fact that technologies I4.0 can be critical to overcoming these weaknesses. A Systematic Literature Review (SLR)-based methodology has been applied to achieve the proposed objectives. SLR provides a structured, rigorous, and objective process to locate, select, and assess existing contributions to a particular study object (Denyer and Tranfield 2009). SLR has allowed the identification and analysis of the extant literature to classify the IT characteristic of I4.0 in LSCM contexts, by their impact on the LSC's responsiveness, specifically, the SC's flexibility and/or response speed.

To achieve these goals, this chapter is structured as follows. The following section presents the background and driving forces. Next, the methodology is explained in detail. The following section describes the findings. The chapter concludes with a section that sets out the main conclusions, gaps, and any lines of future research that have been detected.

6.2 BACKGROUND AND DRIVING FORCES

As occurred in the past, when the steam engine was invented in the 18th century and gave rise to the first industrial revolution, a set of new IT is currently being applied that has great potential for transforming the way that we live, relate to each other, and do business. These IT and their application to business are developing at an even greater rate than could ever have been imagined.

In this context, the Fourth Industrial Revolution, also known as I4.0, pursues the intensive application of some specific IT in production and service environments and in the management of customer and supplier relationships (Kang et al. 2016; Monostori et al. 2016). The intensive application of certain IT, including both mature and fledgeling technologies, aims to create intelligent factories through the full digitization of value flows on both the internal and SC levels (Schmidt et al. 2015; Schrauf and Berttram 2016; Satoglu et al. 2018).

The integration of technologies to process information, robots, intelligent software and sensors, among other things, thus enables a number of capabilities in companies, such as self-configuration, self-adjustment, and self-optimization, which allow more flexible and agile processes and facilitate LSCM (Schmidt et al. 2015; Satoglu et al. 2018; Tortorella and Fettermann 2018). The development of these capabilities spans all the different agents in the LSC from suppliers to customers and enables the information and physical flows involved to be planned, forecast, and controlled intelligently.

Among the main effects of IT in the I4.0 framework are extensive automatic machine interconnection, communications networks, advanced robotics, advanced data processing technology integration, and the ability to situation self-check (Tortorella, Miorando and Mac Cawley 2019). So, I4.0 renders Lean production systems more autonomous, precise, flexible, and dynamic (Tortorella and Fettermann 2018), allowing them to connect with each other and, consequently, enables an LSC to be constructed and improved. In fact, the adoption of some specific facilitating technologies not only entails the digitization of processes, products, and company functions but, also, of SC members (Schrauf and Berttram 2016).

For example, in LSCM environments, I4.0 can facilitate the principle of "pull" planning and enable planning to be carried out with production aligned with real-time demand information. A perfect fit can thus be achieved between real demand and these plans, as well as synchronization – also in real time – of SC members and dynamic self-optimization of the processes and agents involved. Therefore, I4.0 improves responsiveness to changes in demand, capacity, purchases, and supply at the LSC level. And so, for all these reasons, despite incorporating the latest technologies and intelligent algorithms, I4.0 is not incompatible with Lean principles but, rather, the smart factory allows itself to be built on the foundations of LM (Sanders, Elangeswaran and Wulfsberg 2016).

Companies are currently becoming increasingly aware of their growing dependence on their supplier and customer networks and, as a result, on the need to manage and integrate their SCs, from the raw materials supplier to end consumers (Fisher 1997; Lambert, Cooper and Pagh 1998; Croom, Romano and Giannakis 2000; Jack and Raturi 2002). In this sense, a great number of studies show the importance of improving SCs' responsiveness for satisfying customer demands (Fisher et al. 1994; Vickery, Calantone and Dröge 1999; Olhager and West 2002). Responsiveness is connected with flexibility and response speed (Stevenson et al. 2007).

SC responsiveness means the way that a SC is capable of adapting to environmental changes and variable demand characteristics (Reichhart and Holweg 2007; Swafford, Ghosh and Murthy 2008). It can be understood from a dual perspective: a) as an improvement in the SC's flexibility and b) as an improvement in its quickness or speed of response. Flexibility seeks to connect physical and information flows effectively and efficiently from customers to suppliers so as to gain the ability to adapt or to have an adaptive response to both internal and external environmental uncertainty (Vickery, Calantone and Dröge 1999; Vonderembse et al. 2006; Reichhart and Holweg 2007). Response speed is linked to the ability to adapt quickly to environmental changes such as real changes in demand (Christopher and Towill 2000; Gligor, Holcomb and Stank 2013; Qrunfleh and Tarafdar 2013).

Only the IT identified with I4.0 that have been considered in the literature on LSCM are described in the following. The way that these IT act as driving forces of certain LSCM principles and goals is also presented in global terms. However, as has been seen in preceding chapters in this book, the IT associated with I4.0 form an even larger group.

One of the most representative of the I4.0 IT is the Internet of Things (IoT), which consists of sensors and Cyber Physical Systems (CPS) that connect to the Internet and that interact with their internal state and/or the external environment. IoT can be used internally in companies and at the SC level (Abdel-Basset, Manogaran and Mohamed 2018). IoT devices applied at the LSC level afford products visibility and traceability and, even more importantly, they contribute to information integration with customers and suppliers (Xu et al. 2018). IoT has the ability to increase organizations' operating efficiency, which could become a competitive advantage thanks to better identification, tracking, and monitoring of the status of assets throughout the SC and the achievement of Lean's objectives.

Closely linked to IoT is the technology known as Radio Frequency Identification (RFID) systems, which consists of labels with embedded chips that can store information and transfer it over a telematic network by radiofrequency (Ngai et al. 2008). When applied to LSCM, this allows improvements to communication mechanisms between SC members by automating the production, capture, tracking, and transmission of information in real time between customers and suppliers (Angeles 2005; Want 2006).

Another of I4.0's benchmark technologies is Cloud Computing. This consists of a parallel, distributed infrastructure system formed of a large number of widely dispersed computers and storage systems that connect with each other and through which it is possible to aggregate, share, and select virtualized computing resources (Buyya 2009). Cloud Computing allows services to be offered with multiple advantages aligned with the objectives of Lean, such as a reduction in costs, better resource exploitation, the need for smaller direct investments, and greater accessibility, among other things.

With links to Cloud Computing, Big Data involves the identification, capture, processing, analysis, and visualization of an enormous amount of data to support decision-making (LaValle et al. 2011). In Big Data, the use of advanced techniques to analyze enormous volumes of data can help companies to uncover hidden patterns, trends, or customer preferences, among other matters. This would help companies to better adapt to customers and suppliers in LSCM environments by helping to identify environmental changes earlier and to anticipate competitors, and favouring competitiveness (Kwon, Lee and Shin 2014; Gupta, Modgil and Gunasekaran 2019).

Artificial Intelligence (AI) is yet another of I4.0's benchmark technologies that the literature has linked to LSCM. This consists of providing systems with the ability to interpret external data, learn from them, and use them to achieve goals and carry out specific tasks, thus emulating human intelligence (Kaplan and Haenlein 2019). In the LSCM context, the current environment's complexity, dynamism, and uncertainty would seem to make AI a powerful tool to support decision-making and achieve Lean objectives (Liu et al. 2013) in supplier selection, for example, and in any Maintenance, Repair and Overhaul (MRO) activity that involves suppliers and customers.

Virtual Reality (VR) enables the generation of virtual environments that make users feel that they are immersed in real surroundings. It allows stimuli to be perceived, and interaction with the environment by transcending the spatial and physical constraints found in the real world (Jayaram, Connacher and Lyons 1997). In LSCM contexts, VR could enable simulation activities and also help SC members to learn operations management by instructing personnel and providing them with the necessary skills.

Autonomous Vehicles (AV) are also a widely used type of IT in LSCM. Autonomous Carriers (AC) for material flow objects in logistics can be described as the decentralized coordination of intelligent logistics objects (e.g., transport objects and means, products, machines) and the routing through a logistics system by the intelligent parts themselves (Scholz-Reiter and Freitag 2007). In Lean contexts applied to production and logistics environments, AV can contribute to the continuous flow of materials in SCs by automating this activity, reducing errors, and synchronizing suppliers and customers, whilst at the same time bringing down costs.

Additive Manufacturing or 3D Printing is an I4.0 technology that enables additive layering to turn a digital design in a file into a good. The materials used include polymers, resin, and metal, among others. When applied in an LSCM environment, AM could enable on-demand production at exactly the right time, one-piece-flow, and substantial product customization (Tziantopoulos et al. 2019).

Blockchain is a Cloud infrastructure technology that processes data in such a way that tamper-proof recordings are made of any information and transactions owing to the information being stored in a distributed and decentralized infrastructure and highly replicated (Perboli et al. 2018; Treiblmaier 2018). Any attempt to alter the information is detected and it is rebuilt, and so cannot be altered. Although Blockchain emerged as a leading technology for financial applications, recently it is considered as a disruptive IT for structuring a new digital SC based on decentralization, secure transactions, and real-time updating (Perboli et al. 2018). This technology's influence on LSCM is related to information traceability, visibility, and high security to protect against any attempts to alter information (Field 2017; Kshetri 2017).

6.3 METHODOLOGY

To obtain an overview of the current state of knowledge and of the aspects and implications that exist between I4.0 and LSCM, the most relevant contributions have had to be identified in the extant literature in the area that is the object of study. For this, the process proposed for a Systematic Literature Review (SLR) has been applied (Tranfield, Denyer and Smart 2003; Denyer and Tranfield 2009; Thomé, Scavarda and Scavarda 2016; Durach, Kembro and Wieland 2017). This methodology enables the current state of existing research into any particular topic to be understood through the application of a structured, rigorous, and objective process. When the proposed SLR process is correctly applied, the most relevant contributions in an area of study can be identified and the existing information regarding a specific research question synthesized, thus providing a general overview of the current state of research (Tranfield, Denyer and Smart 2003; Denyer and Tranfield 2009).

The five proposed stages for an SLR (Denyer and Tranfield 2009) have been followed in this study in conjunction with the recommendations and proposed best practices for its specific application in Operations Management in general (Thomé, Scavarda and Scavarda 2016) and the SC in particular (Durach, Kembro and Wieland 2017). In the process that is followed, the research is got underway by establishing a research protocol that details specific issues that have to be taken into account as the research is developed. So, a series of guidelines are laid down to prevent any errors and to guarantee consistency and coherence when a number of different researchers are involved (Thomé, Scavarda and Scavarda 2016). This protocol contains the procedure that should be adhered to, to achieve the proposed goals and includes issues related to search strategies, criteria for including and excluding articles, criteria for identifying contributions, and details for coding the information extracted from the articles.

The specific steps followed were taken from (Tranfield, Denyer and Smart 2003; Denyer and Tranfield 2009), as shown in Figure 6.1: (1) research question formulation; (2) locating the literature; (3) study selection and evaluation; (4) analysis and synthesis of results; (5) reporting of results.

With the scope of the study delimited and the research goals specified (1), the literature is located (2). For this, the Web of Science (WoS), Scopus, and ABI Inform databases have been used. The search chains have been designed as combinations of three blocks of words using simple and Boolean operators. The area that is the object of study is the intersect between I4.0, Lean, and SC. The search keywords used are given in Table 6.1. The syntax of the search chains has been adapted to each of the databases used.

The SLR process then continues (3) with the selection and evaluation of the studies identified in the previous step. The objective at this stage is to eliminate any works that appear in the searches, but which are not relevant and/or are unrelated to the object of study (false positives). For this, inclusion and exclusion criteria have been established that determine which studies are to be considered

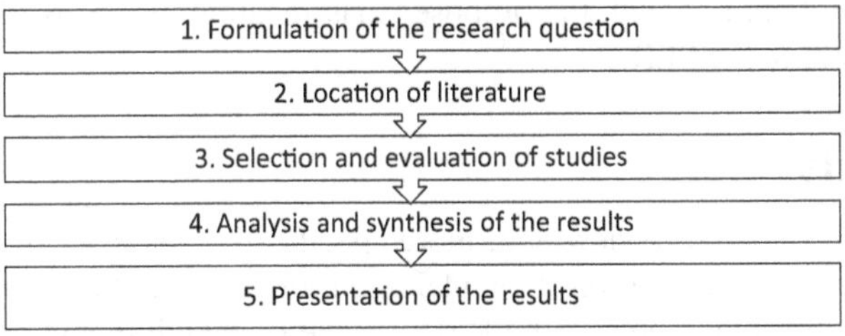

| 1. Formulation of the research question |
| 2. Location of literature |
| 3. Selection and evaluation of studies |
| 4. Analysis and synthesis of the results |
| 5. Presentation of the results |

FIGURE 6.1 Systematic Literature Review (SLR): stages: (Adapted from Denyer and Tranfield 2009).

TABLE 6.1

Search Keywords

Lean: Lean, JIT, Just-in-Time

Supply Chain: supply chain, logistics

Industry 4.0: Industry 4.0, information system, information technology, information and communication technology, ICT, technological innovation, internet of things, IoT, cloud, web, e-business, e-commerce, radiofrequency identification, RFID, business intelligence, virtual reality, augmented reality, robot, artificial intelligence, big data, blockchain, autonomous carriers

in the research, and those that are to be discarded with no additional analysis. To ensure the relevance of the selected works, only articles and press articles published in English between 1996 (first article on LSCM, Lamming 1996) and December 2018 have been considered. The search has been restricted to the areas of research included in the scope of this study. Subsequently, the articles have been evaluated by reading the titles, abstracts, and keywords. Any articles that clearly do not comply with the research question have been discarded, and in cases where doubts have arisen, the theoretical framework and the main results of the article have been read. Furthermore, during this stage, a check has been made to ensure that the selected articles have been published in journals indexed in *Journal Citation Reports (JCR)* and/or *Scimago Journal Rank (SJR)* to guarantee the peer review process and the quality of the articles. This process yielded a final selection of 23 articles.

In (4) the analysis and synthesis of the results, each of the selected works has been read in its entirety. As this stage has been conducted by more than one researcher, as specified in the research protocol (Thomé, Scavarda and Scavarda 2016), a database has been constructed in Excel with the main ideas, objectives, and contributions of each of the analyzed works to ensure the consistency of the results and enable their interpretation. The main technologies involved have also been identified, as has other interesting complementary information such as the title, author, journal, year of publication, methodology, and journal impact.

The last stage of an SLR consists of (5) reporting the end results of the analysis and synthesis. The following section describes this stage in detail. First, a classification of the existing literature is proposed, followed by the presentation and discussion of the results in line with the proposed classification. Figure 6.2 illustrates the process followed in the SLR.

6.4   RESULTS

The systematic literature review and, more specifically, the analysis and synthesis process have enabled a criterion to be identified for grouping the identified articles by the contributions that the specific I4.0 IT make to flexibility and response speed in LSCM contexts. Table 6.2 gives the novel classification of the articles based on this classification criterion.

This classification has enabled the identification of the technologies that contribute flexibility to the LSC that have been the object of attention in the literature. These are IoT, RFID, Cloud Computing, and VR, with Cloud Computing being the technology that has received the most attention from researchers in recent years. With respect to the technologies that contribute to response speed most developed in the literature in the context of LSC, these were IoT, RFID, and AI. On the other hand, technologies such as Blockchain, Big Data, Autonomous Vehicles, and Additive Manufacturing have not been greatly developed in LSCM research. There is, moreover, a consensus among researchers that technologies such as IoT, RFID, and Cloud Computing contribute to both flexibility and response speed in the LSC.

Described in greater detail below are, first, the IT that affect flexibility in LSCM environments and, second, the IT that affect response speed.

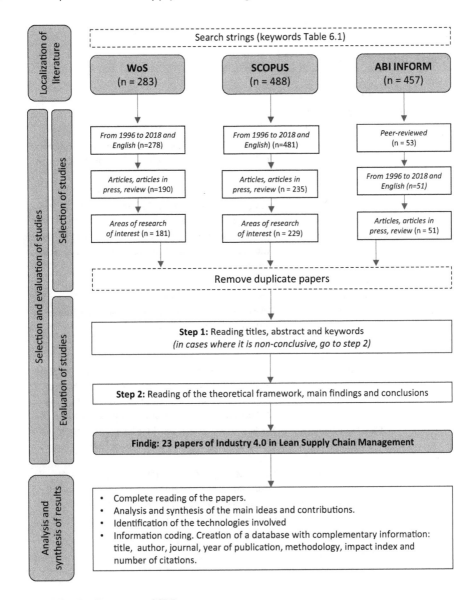

FIGURE 6.2   Summary of SLR.

6.4.1   Influence of Industry 4.0 on LSCM Flexibility

The IT analyzed in the literature that can potentially contribute to improvements to flexibility in the LSCM framework are: (1) IoT; (2) RFID; (3) Cloud Computing; (4) VR; (5) Blockchain. Described below are the roles played by each in this respect.

6.4.1.1   Internet of Things (IoT)

IoT contributes to LSCM flexibility by enabling the integration of the different processes that generate value throughout the SC. It does this by connecting and

TABLE 6.2

Classification of Articles

Group (second classification level)	Line of research (first level of classification)	Articles
Flexibility	Internet of Things	Dave et al. (2016); Sanders, Elangeswaran and Wulfsberg (2016); Hofmann and Rüsch (2017); Yerpude and Singhal (2017)
	Radio Frequency Identification	Otamendi, García-Higuera and García-Ansola (2011); Saygin and Sarangapani (2011); Shin et al. (2011); Powell and Skjelstad (2012)
	Cloud Computing	Sanders, Elangeswaran and Wulfsberg (2016); Hofmann and Rüsch (2017); Vazquez-Martinez et al. (2018); Xu et al. (2018)
	Virtual Reality	Li et al. (2018)
	Blockchain	Perboli, Musso and Rosano (2018)
Response speed	Internet of Things	Dave et al. (2016); Sanders, Elangeswaran and Wulfsberg (2016); Hofmann and Rüsch (2017); Yerpude and Singhal (2017); Xu et al. (2018)
	Radio Frequency Identification	Otamendi, García-Higuera and García-Ansola (2011); Saygin and Sarangapani (2011); Dai et al. (2012); Huang et al. (2012); Powell and Skjelstad (2012); Zelbst et al. (2014); Sanders, Elangeswaran and Wulfsberg (2016); Nabelsi and Gagnon (2017); Tsao, Linh and Lu (2017)
	Cloud Computing	Sanders, Elangeswaran and Wulfsberg (2016); Hofmann and Rüsch (2017); Vazquez-Martinez et al. (2018); Xu et al. (2018)
	Artificial Intelligence	Güner, Murat and Chinnam (2012); Liu et al. (2013); Hofmann and Rüsch (2017)
	Autonomous Vehicles	Mehrsai, Thoben and Scholz-Reiter (2014)
	Big Data	Christopher and Ryals (2014)
	Additive Manufacturing	Christopher and Ryals (2014)

exchanging information, mainly by the technology providing improvements to interoperability between the main information systems (Dave et al. 2016; Sanders, Elangeswaran and Wulfsberg 2016). Information flows generated between SC members as a result of IoT implementation facilitate production planning and control (Dave et al. 2016; Hofmann and Rüsch 2017; Yerpude and Singhal 2017) by a precise deduction of demand patterns in real time and, in the final instance, allowing the Lean systems in an LSC to respond more flexibly to evolving customer demands and market turbulences (Hofmann and Rüsch 2017; Yerpude and Singhal 2017). The benefits that IoT brings enable the medium- and long-term development of close ties between customers and suppliers, mostly as a result of IoT offering greater transparency in the LSC (Dave et al. 2016; Yerpude and Singhal 2017). Having real data on processes and products at all times means that these can be better planned, and actions can be taken to meet customer needs in the medium or long term (Dave et al. 2016; Yerpude and Singhal 2017). In this regard, Yerpude and Singhal (2017) highlight that Vendor Managed Inventory (VMI) enabled with IoT helps the Lean systems to track inventory in real time (e.g., categorizing it into Fast, Slow and Non-moving, FSN) and to achieve a more flexible SC through an improved replenish pull system (Sanders, Elangeswaran and Wulfsberg 2016).

6.4.1.2   Radio Frequency Identification (RFID)

RFID technologies are very closely related to IoT and are used in LSCM to handle information automatically (Otamendi, García-Higuera and García-Ansola 2011; Powel and Skjelstad 2012). So, the creation of information management platforms with RFID technology support enables the development of an efficient management framework capable of perfectly integrating the heterogeneous and dynamic environments of the various SC agents. This is an effective support for information exchange and gives the LSC added flexibility (Shin et al. 2011). Furthermore, RFID has a strong impact on stock control by improving the monitoring of material flows across the LSC (Saygin and Sarangapani 2011; Powel and Skjelstad 2012) thanks to the greater visibility given to products and their associated information. This benefits organizations, as they are able to collaborate, plan, supervise, and execute whilst continuously improving the operating features in the value flow (Saygin and Sarangapani 2011; Powel and Skjelstad 2012). In this line, RFID helps boost LSC flexibility, trust, and security by eliminating errors in the capture and processing of product-related information and thus supports the ongoing improvement of an organization's processes (Otamendi, García-Higuera and García-Ansola 2011; Powel and Skjelstad 2012). Thus, RFID systems in LSCM environments have contributed to a better and flexible decision-making process when the initial plan needs to be changed and a more flexible response to changes (Shin et al. 2011) by leveraging LSCM principles (Saygin and Sarangapani 2011) and some backbone techniques such as Value Stream Mapping (Powel and Skjelstad 2012).

6.4.1.3   Cloud Computing

When implemented in LSCM, Cloud technologies allow the flexible integration of different systems and technologies (Sanders, Elangeswaran and Wulfsberg 2016; Hofmann and Rüsch 2017; Xu et al. 2018). Many authors agree on highlighting Cloud technologies' ability to enable a framework of collaboration between SC members, which results in greater SC synchronization/integration and effective feedback between suppliers and customers (Sanders, Elangeswaran and Wulfsberg 2016; Hofmann and Rüsch 2017; Vazquez-Martinez et al. 2018; Xu et al. 2018). Thanks to its all-pervasiveness and interoperability, Cloud Computing enables data to be collected and handled throughout the entire SC. Heterogeneous assets can be managed and information exchanged between customers and suppliers in such a way that they can share and act on the same information flexibly and at a minimum cost, whilst the SC's transparency and flexibility are increased in such a way as to prevent or reduce the bullwhip effect (Hofmann and Rüsch 2017; Xu et al. 2018). This is thanks to all connected agents that would be able to act on real time. In fact, real time consumption of products may automatically trigger supply orders or inform suppliers and, therefore, they may gain additional flexibility (Hofmann and Rüsch 2017). In addition, Cloud Computing makes a shared economy possible, reducing both costs and the need for investments in infrastructure and enabling the use of other complementary technologies in small- and medium-size companies such as IoT, which provides configurable and scalable information services along a SC and, ultimately, enables business processes to be managed more flexibly, efficiently, and economically (Xu et al. 2018).

6.4.1.4   Virtual Reality

When applied in LSCM, VR is closely linked to learning processes in very complicated procedures. So, in Prefabrication Housing Production, for example, it enables the site assembly process and logistics processes to be replicated (Li et al. 2018). And it has proven its utility by providing the agents involved in the process with the necessary skills and an understanding of how processes are executed and familiarizing them with the computer programs used. Training can be done anywhere with consequent savings in space, time, and money (Li et al. 2018). So, in the final instance, VR contributes to increasing the flexibility of the companies that use it. Thus, VR enables the greater adaptation of workers and SC members to new processes, mitigating any possible uncertainties, detecting constraints, reducing and/or eliminating any possible errors and, ultimately, improving the decision-making process (e.g., assessment of production quality, remaining time to the site assembly, lead time of responding to changes) and optimizing processes throughout the LSC (Li et al. 2018).

6.4.1.5   Blockchain

Thanks to its ability to guarantee the inalterability, integrity, and scalability of information, in LSCM contexts, Blockchain builds greater trust between SC members to the benefit of interoperability. The heterogeneous and dynamic nature of the information generated by customers and suppliers throughout the SC and the difficulty of integrating the information systems of all the organizations involved could be mitigated by this technology. If Blockchain is adopted by customers and suppliers, it could guarantee organizations information integration and interoperability with their existing systems (Perboli, Musso and Rosano 2018). Trust between members reduces the need for additional checks and provides for common, inalterable information that confers on organizations a greater medium- or long-term capacity for planning and adaptation to address environmental changes. This greater flexibility derives from the transparency and visibility of information that the technology offers in LSCM environments. Inaccurate product demand and flow information have short-term negative effects on both customers and suppliers. However, when the entire SC is visible (inbound and outbound processes), manufacturers can optimize production processes in the medium and long term, anticipate any possible fluctuations in demand, and reduce the bullwhip effect, and thus enjoy greater flexibility. In this way, inventories in LSCM environments that are currently maintained at high levels higher to increase protection against the bullwhip effect could be reduced and this would result in a reduction in overall logistics costs and higher margins and profitability (Perboli, Musso and Rosano 2018). The use of Blockchain would, therefore, enhance an LSC's flexibility. Furthermore, organizations are generally reluctant to share sensitive information with their customers and suppliers. However, using this technology in LSCM contexts would improve security levels in information exchange and, consequently, mitigate these concerns. This would benefit collaborative environments and information sharing and thus drive up the SC's efficiency (Perboli, Musso and Rosano 2018).

6.4.2 Influence of Industry 4.0 on the Response Speed of LSCM

The technologies analyzed in the literature that can potentially contribute to response speed in LSCM contexts are: (1) IoT; (2) RFID; (3) Cloud Computing; (4) AI; (5) AV; (6) Big Data; (7) Additive Manufacturing. The role that these IT play in enhancing the SC's response speed in LSCM contexts, according to the analyzed literature, is described below.

6.4.2.1 Internet of Things (IoT)

IoT has sufficient potential to impact almost all sectors, from retail to the automotive industry. The contributions that its implementation makes in LSCM contexts are related to significant improvements in capacity and rapid adaptation to changes in demand due to greater anticipation and faster response to unforeseeable situations. IoT enables data to be collected and analyzed automatically, as well as managed remotely, by providing real-time visibility, traceability, and control of SC processes (Xu et al. 2018). When focal companies, suppliers, and customers implement this IT, it provides a substantial opportunity to generate constant information flows (continuous data vs. discrete data) and to fully or partially automate several communication functions across the SC (Dave et al. 2016). It is then possible to monitor not only the progress of processes and the flow of physical elements in real time but also the information associated with them, enabling the Lean principle of visual management through IoT devices (Dave et al. 2016; Sanders, Elangeswaran and Wulfsberg 2016). It makes physical and information flows visible at key points in the SC at all times, resulting in real-time monitoring and control of stock levels and precise location (Hofmann and Rüsch 2017). The generation of information flows and real-time data transmission between the focal company and its suppliers and customers enable much more effective and efficient short-term planning (Yerpude and Singhal 2017). All of this enables reductions in inventory levels, makes inventory easier to manage, and helps reduce response times and lead times, thus significantly contributing to improving the SC's response time (Hofmann and Rüsch 2017; Yerpude and Singhal 2017).

6.4.2.2 Radio Frequency Identification (RFID)

RFID is an IT that is very closely linked to IoT. It enables objects to be identified with no human intervention, thus giving real-time visibility to physical flows along the SC thanks to the accurate and detailed information that the technology provides (Saygin and Sarangapani 2011; Powel and Skjelstad 2012). RFID systems have contributed to improving the LSC's response speed thanks to faster information capture and transmission in conjunction with greater information integration. It, therefore, affords the SC new capabilities for a much more agile flow of products, whether goods or services, such as in the case of automatic picking, for example. So, in LSCM contexts, RFID has a knock-on effect on response speed to address frequent market changes and reduce the time needed for product movement and distribution (Otamendi, García-Higuera and García-Ansola 2011). Thus, for instance, Just-in-Time and the constant visibility of information provided by RFID systems allow SC members to reduce waiting and processing times (waste), to improve inventory management, to leverage the continuous flow principle (Saygin and Sarangapani 2011) and, in the final instance, to improve the LSC's response speed (Otamendi, García-Higuera and García-Ansola 2011). In addition, the opportunities that RFID systems offer for information to be quickly and autonomously captured in LSCM environments have enabled the automation of processes such as the monitoring and intelligent reallocation of orders, which make Just-in-Time and Just-in-Sequence significantly easier (Shin et al. 2011; Nabelsi and Gagnon 2017; Sanders, Elangeswaran and Wulfsberg 2016; Tsao, Linh and Lu 2017). Improvements to the real-time traceability and visibility of physical flows provided by RFID use have resulted in better inventory control, with reductions in customers' and suppliers' lead times and the time associated with inventory-related decision-making –e.g., in case of suppliers' delays (Dai et al. 2012; Huang et al. 2012; Zelbst et al. 2014; Moon et al. 2018). All the above means that an LSC can operate with greater agility and speed.

6.4.2.3 Cloud Computing

Cloud Computing represents a leap forward from the traditional communication mechanisms between the various SC members. These systems and the use of intelligent devices such as smartphones and tablets enable faster access to information anytime, anywhere (Sanders, Elangeswaran and Wulfsberg 2016). In contexts where LSCM has been adopted, Cloud Computing contributes to response speed, which enables large volumes of data to be stored and processes to be monitored in real time, with all SC members able to access information anytime, anywhere (Vazquez-Martinez et al. 2018; Xu et al. 2018). Customers and suppliers can use Cloud Computing to manage heterogeneous assets and exchange information, with the agents involved able to share and act upon the same information in real time. This improves communication and cooperation between them and, in short, improves the time needed to respond to any changes that might occur in the environment (Hofmann and Rüsch 2017; Xu et al. 2018). Specifically, Vazquez-Martinez et al. (2018) found that certain Cloud solutions could even reduce the negotiation to a partner-to-partner scheme, which translates into a minimization of coordination and synchronization requirements to support the data exchange among partners and, therefore, into a rapid response to changes.

6.4.2.4 Artificial Intelligence (AI)

The current environment in which organizations operate is characterized by complexity, dynamism, and uncertainty. These characteristics are even more marked in the context of the SC. In this sense, organizations are developing powerful AI-based support systems for decision-making throughout the SC (Liu et al. 2013). So, for example, AI-based systems are used to intelligently map routes and these systems have great potential for improving deliveries in LSCM environments such as Just-in-Time and Just-in-Sequence (Güner, Murat and Chinnam 2012; Hofmann and Rüsch 2017). Intelligent routing systems based on real-time traffic information have enabled improvements to delivery speeds throughout the SC, reductions in lead times, and greater customer satisfaction (Güner, Murat and Chinnam 2012). They have also allowed appropriate transport to be selected based on specific order requirements (quantities, delivery dates, etc.), and so the use of these systems means that transportation processes are no longer planned individually but in an integrated way throughout the entire SC (Hofmann and Rüsch 2017). In objective criteria and knowledge-based decision-making processes, the use of AI-based systems is enabling the elimination of any type of activity that does not add value along an LSC. Any wastage, such as excess production and inventory, can be eliminated, whilst lead times and the unnecessary movement of materials and products can be reduced. All of this has a direct impact on the LSC's response speed and, consequently, drives up profitability and productivity (Liu et al. 2013).

6.4.2.5   Autonomous Vehicles (AV) – Autonomous Carriers (AC)

The literature has analyzed the use of a specific type of autonomous vehicle in LSCM contexts: Autonomous Carriers (AC) automatically distribute materials to production chains and thus involve the suppliers of these materials. Vehicles of this type contribute to improving the SC's response time and their use is associated with rapid decision-making for smart individual entities and fast operations, as components and raw materials are supplied automatically. In addition, logistics process times are shortened, as the movement rate is automatic and the vehicles identify local obstacles and manage them by responding to them (self decision-making and control), which enables logistics process times to be substantially reduced. In short, the use of these systems in LSCM results in increased performance, effectiveness, efficiency, and faster response speed and reduced lead times throughout the SC (Mehrsai, Thoben and Scholz-Reiter 2014). A broad range of simulated experiments using mathematical calculations have amply proven AC's capabilities and their use has led to improvements in the required response capability and performance increases in dynamic circumstances and real-time work. AC have, therefore, been seen to contribute directly to raising the response speed in LSCM contexts by permitting a continuous material flow, shorter processing time in logistics operations, reduced lead times and work in progress, greater machine utilization and the avoidance of idleness in queues (waste) (Mehrsai, Thoben and Scholz-Reiter 2014).

6.4.2.6   Big Data

The application of new technologies such as Big Data is transforming SCs, with major changes from what we know today. In a sustainable world, the SC needs to be designed from the customer backwards (demand pull) instead of from the factory outward (supply push), making it responsive to customer demands and reducing waste and returns (Christopher and Ryals 2014). Emerging new technologies such as Big Data enable huge volumes of customer data to be processed and their needs anticipated, with demand identified much more accurately (Christopher and Ryals 2014). Processing large amounts of customer data also enables the value proposal to be better specified through the Lean lens and extending Big Data usage to SCs allows these to operate with less inventory and the SC as a whole to respond more rapidly to any changes demanded by customers (Christopher and Ryals 2014). As a result, Big Data enables changes in consumer preferences to be identified and any need for new products or opportunities in new markets to be identified. In LSCM contexts, Big Data will enable faster decision-making based on information analysis and will be aligned with any changes in the environment, thus facilitating a real-time response to address these changes (Christopher and Ryals 2014). So, in LSCM environments these technologies will enable the development of dynamic capabilities that will allow companies to rapidly respond to any changes in the environment (Christopher and Ryals 2014).

6.4.2.7   Additive Manufacturing

Other manufacturing technologies that are emerging in Industry 4.0, such as additive manufacturing/3D printing, have also been analyzed by the extant literature in LSCM contexts. Additive manufacturing has been found to shorten product development, production, and commercialization times, thus improving short-term response time (Christopher and Ryals 2014). In this sense, digital designs enable additive manufacturing to postpone production to the very last possible moment in the SC (Christopher and Ryals 2014). This reduces any wastage linked to the unnecessary movement of goods and so shortens the times associated with materials supply and increases speed in Just-in-Time environments. At the same time, AM enables small-footprint manufacturing changing the traditional paradigm of the "economies of scale" to "economies of scope" and, ultimately, allows the ability to achieve mass-customization in LSCM environments. Another advantage of additive manufacturing in LSCM environments is the possibility of simplifying the generation of prototypes, with substantial reductions in the times associated with this activity and also reductions in production time and costs associated with products (Christopher and Ryals 2014). As occurs with Big Data, in LSCM contexts, these technologies enable dynamic capabilities to be developed that improve the response speed to environmental changes (Christopher and Ryals 2014).

6.5   CONCLUDING REMARKS

In this work, we have reviewed the literature on the role played by I4.0 IT in LSCM contexts. For this, we have conducted a Systematic Literature Review that has enabled us to identify existing articles of interest and to subsequently propose their classification according to the pursued goal of improving an LSC's responsiveness capability (flexibility and/or response speed) and to the specific IT used in an LSCM framework. This novel classification has enabled us to detect the contributions that specific technologies make to the SC's responsiveness capability in terms of flexibility and/or response speed and the implications that these technologies have in LSCM environments.

As has been shown in this chapter, in an LSCM context I4.0 is underpinned by tools and IT that improve both flexibility and response speed in the relationships formed between focal companies and their customers and suppliers throughout the SC. Thereby, I4.0 IT are critical in overcoming one the main weaknesses of LSCM environments (poor mass-customization capability and limited adaptation to changing environmental requirements), enabling highly flexible mass production systems, interconnected in real time, and leveraging some cornerstones of LSCM such as the Pull principle (e.g., Sanders, Elangeswaran and Wulfsberg 2016; Hofmann and Rüsch 2017).

So, on the one hand, the literature shows that organizations that develop and implement I4.0 IT in LSCM environments can be more innovative, are able to identify changes in consumer preferences before their competitors do, and detect the need for new products and opportunities in new markets. This enables firms to react and adapt to these changes in the medium and long term or, to put it another way, to increase their flexibility. However, on the other hand, I4.0 in LSCM environments enables dynamic capabilities to be developed that allow customers and suppliers to rapidly adapt to changes in the environment. They are able to respond to these changes with agility in the short term by accelerating decision-making thanks to rapid information analysis and immediate response, with an impact throughout the LSC. In other words, it produces a general increase in response speed in LSCM environments.

Furthermore, when implemented in LSCM, I4.0 provides the necessary mechanisms and capabilities that foster collaboration environments and the integration of customers and suppliers throughout the SC, with a direct impact on flexibility. It also allows the logistics, production, and commercialization processes in which a wide range of agents intervene to be stepped up and this has a direct impact on response speed throughout the entire SC.

According to the development and content of this chapter, customers and suppliers need to stay constantly informed about the status of goods and services if they are to achieve the goals sought by LSCM. Companies can then address changes in demand and consumer wishes and preferences (flexibility) with the least possible wastage and take action more quickly (response speed). In this sense, when properly implemented and used by customers and suppliers, the joint use of I4.0 IT such as IoT, Cloud Computing, and Artificial Intelligence in LSCM contexts enables immediate, automatic feedback between customers and suppliers and thus overcomes the paperwork inherent in inadequate communication channels and achieves the sought-after Lean objectives throughout the entire SC.

The growing volume of information generated in companies and SCs requires more developed advanced analytical tools such as Big Data. Moreover, the combination of Artificial Intelligence, Cloud technologies, and Big Data make these tools available to all SC members, facilitating decentralized decision-making based on objective criteria. The increased volume of information flowing between LSC members and its sensitive and confidential nature require strict security measures. So, Blockchain could provide the required security and be of use for monitoring high-value products during logistics processes.

In the LSCM context, additive manufacturing and 3D printers would enable production to be delayed until the very last moment in the SC. These technologies give the opportunity to simplify prototype generation, reduce production times and costs, and contribute to the SC's response time, adaptability, and competitiveness.

In relation to possible future lines of research, some IT have been detected which, despite playing a major role in I4.0, have not been analyzed in LSCM contexts. This is the case in Augmented Reality and Robotics (Cobots). In other cases, the extant research on other benchmark I4.0 IT is very superficial and scanty. This is the case in technologies such as Big Data, Virtual Reality, and Blockchain.

An IT such as Augmented Reality, which is a major feature in I4.0 that enables virtual information to be added to physical objects, has not been addressed by researchers in the context of LSCM. Augmented reality could have a major impact on achieving the LSCM objectives of locating goods and streamlining delivery if applied in warehousing or distribution, as in both cases it would enable goods to be located more easily, more efficiently, and more rapidly. Therefore, its impact on the fastest short-term response (response speed) could be powerful in LSCM environments. For example, the use of Augmented Reality could enable a line on the warehouse floor to be visualized in order to rapidly locate a target. Its application would be extremely beneficial to last mile-related logistics processes and would drastically reduce the time carriers take to locate goods in delivery vehicles. In the case of technologies such as robotics, in the LSCM context, these could be applied to repetitive tasks such as picking and packing merchandise, and so could increase the speed and accuracy of routine operations. In addition, there could be gains in efficiency through their joint work with humans (autonomatization) and reduced risk of injuries to workers in hazardous environments. It would, therefore, be useful to investigate the role that it plays, both for efficiency and taking prompt action as, as has been indicated, none of these issues have been examined in LSCM contexts and they should be addressed in the near future.

There are some limitations to this study. Some specific keywords were used to locate the literature in the WoS, Scopus, and ABI Inform databases. The choice of these words is always complicated and rather subjective. The classification also corresponds to criteria determined by the researcher, and this again entails a degree of subjectivity. Both are limitations inherent in SLR methodology. Nevertheless, despite the mentioned limitations, we, the authors, believe that this study has provided some interesting findings that help contextualize the role of I4.0 IT in delivering SC responsiveness in LSCM environments.

ACKNOWLEDGMENTS

The authors are grateful for the financial support for this study from Research Project ECO2015-65874-P, funded by the Ministry of the Economy and Competitiveness.

REFERENCES

Christopher, M. and L. J. Ryals. 2014. The supply chain becomes the demand chain. *Journal of Business Logistics*, 35(1):29–35.
Dai, Q., R. Zhong, G. Q. Huang, T. Qu, T. Zhang and T. Y. Luo. 2012. Radio frequency identification-enabled real-time manufacturing execution system: A case study in an automotive part manufacturer. *International Journal of Computer Integrated Manufacturing*, 25(1):51–65.

Dave, B., S. Kubler, K. Främling and L. Koskela. 2016. Opportunities for enhanced lean construction management using Internet of Things standards. *Automation in Construction*, 61:86–97.

Güner, A. R., A. Murat and R. B. Chinnam. 2012. Dynamic routing under recurrent and non-recurrent congestion using real-time ITS information. *Computers & Operations Research*, 39(2):358–373.

Hofmann, E. and M. Rüsch. 2017. Industry 4.0 and the current status as well as future prospects on logistics. *Computers in Industry*, 89:23–34.

Huang G. Q., T. Qu, Y. F. Zhang and H. D. Yang. 2012. RFID-enabled product-service system for automotive part and accessory manufacturing alliances. *International Journal of Production Research*, 50(14):3821–3840.

Li, X., G. Q. Shen, P. Wu, H. Fan, H. Wu and Y. Teng. 2018. RBL-PHP: Simulation of lean construction and information technologies for prefabrication housing production. *Journal of Management in Engineering*, 34(2):1–18.

Liu, S., M. Leat, J. Moizer, P. Megicks and D. Kasturiratne. 2013. A decision-focused knowledge management framework to support collaborative decision making for lean supply chain management. *International Journal of Production Research*, 51(7):2123–2137.

Mehrsai, A., K. D. Thoben and B. Scholz-Reiter. 2014. Bridging lean to agile production logistics using autonomous carriers in pull flow. *International Journal of Production Research*, 52(16):4711–4730.

Moon, S., S. Xu, L. Hou, C. Wu, X. Wang and V. W. Y. Tam. 2018. RFID-aided tracking system to improve work efficiency of scaffold supplier: Stock management in Australasian supply chain. *Journal of Construction Engineering and Management*, 144(2), 1–9.

Nabelsi, V. and S. Gagnon. 2017. Information technology strategy for a patient-oriented, lean, and agile integration of hospital pharmacy and medical equipment supply chains. *International Journal of Production Research*, 55(14):3929–3945.

Otamendi, F. J., A. García-Higuera and P. García-Ansola. 2011. New business opportunities along the air tourism supply chain: The combination of identification technologies and just-in-time. *African Journal of Business Management*, 5(11):4007–4022.

Perboli, G., S. Musso and M. Rosano. 2018. Blockchain in logistics and supply chain: A lean approach for designing real-world use cases. *IEEE Access*, 6:62018–62028.

Powell, D. and L. Skjelstad. 2012. RFID for the extended lean enterprise. *International Journal of Lean Six Sigma*, 3(3):172–186.

Sanders, A., C. Elangeswaran and J. Wulfsberg. 2016. Industry 4.0 implies lean manufacturing: Research activities in industry 4.0 function as enablers for lean manufacturing. *Journal of Industrial Engineering and Management*, 9(3):811–833.

Saygin, C. and C. Sarangapani. 2011. Radio Frequency Identification (RFID) enabling lean manufacturing. *International Journal of Manufacturing Research*, 6(4):321–336.

Shin, T. H., S. Chin, S. W. Yoon and S. W. Kwon. 2011. A service-oriented integrated information framework for RFID/WSN-based intelligent construction supply chain management. *Automation in Construction*, 20(6):706–715.

Tsao, Y. C., V. T. Linh and J. C. Lu. 2017. Closed-loop supply chain network designs considering RFID adoption. *Computers & Industrial Engineering*, 113:716–726.

Vazquez-Martinez, G. A., J. L. Gonzalez-Compean, V. J. Sosa-Sosa, M. Morales-Sandoval and J. C. Perez. 2018. CloudChain: A novel distribution model for digital products based on supply chain principles. *International Journal of Information Management*, 39:90–103.

Xu, G., M. Li, C. H. Chen and Y. Wei. 2018. Cloud asset-enabled integrated IoT platform for lean prefabricated construction. *Automation in Construction*, 93:123–134.

Yerpude, S. and T. K. Singhal. 2017. Augmentation of effectiveness of Vendor Managed Inventory (VMI) operations with IoT data – A research perspective. *International Journal of Applied Business and Economic Research*, 15(16):489–502.

Zelbst, P. J., K. W. Green Jr., V. E. Sower and R. D. Abshire. 2014. Impact of RFID and information sharing on JIT, TQM and operational performance. *Management Research Review*, 37(11):970–989.

COMPLIMENTARY REFERENCES

Abdel-Basset, M., G. Manogaran and M. Mohamed. 2018. Internet of Things (IoT) and its impact on supply chain: A framework for building smart, secure and efficient systems. *Future Generation Computer Systems*, 86:614–628.

Angeles, R. 2005. RFID technologies: Supply-Chain applications and implementation issues. *Information Systems Management*, 22(1):51–65.

Buer, S. V., J. O. Strandhagen and F. T. S. Chan. 2018. The link between Industry 4.0 and lean manufacturing: Mapping current research and establishing a research agenda. *International Journal of Production Research*, 56(8):2924–2940.

Buyya, R., C. S. Yeo, S. Venugopal, J. Broberg and I. Brandic. 2009. Cloud computing and emerging IT platforms: Vision, hype, and reality for delivering computing as the 5th utility. *Future Generation Computer Systems*, 25(6):599–616.

Carr, N. G. *Does IT Matter? Information Technology and the Corrosion of Competitive Advantage*. Boston, MA: Harvard Business School Press, 2004.

Christopher, M. and D. R. Towill. 2000. Supply chain migration from lean and functional to agile and customized. *Supply Chain Management: An International Journal*, 5(4):206–213.

Croom, S., P. Romano and M. Giannakis. 2000. Supply chain management: An analytical framework for critical literature review. *European Journal of Purchasing & Supply Management*, 6(1):67–83.

Denyer, D. and D. Tranfield. Producing a systematic review. In *The Sage Handbook of Organizational Research Methods*, edited by D. A. Buchanan and A. Bryman, 671–689. London: Sage Publications, 2009.

Durach, C. F., J. Kembro and A. Wieland. 2017. A new paradigm for systematic literature reviews in supply chain management. *Journal of Supply Chain Management*, 53(4):67–85.

Field, A. M. 2017. Blockchain for freight? *Journal of Commerce*, 18(5):88–92.

Fisher, M. L. 1997. What is the right supply chain for your product? *Harvard Business Review*, 75(2): 105–116.

Fisher, M. L., J. H. Hammond, W. R. Obermeyer and A. Raman. 1994. Making supply meet demand in an uncertain world. *Harvard Business Review*, 72(3): 83–93.

Frohlich, M. T. and R. Westbrook. 2001. Arcs of integration: An international study of supply chain strategies. *Journal of Operations Management*, 19(2):185–200.

Gligor, D. M., M. C. Holcomb and T. P. Stank. 2013. A multidisciplinary approach to supply chain agility: Conceptualization and scale development. *Journal of Business Logistics*, 34(2):94–108.

Gunasekaran, A., K. H. Lai and T. C. E. Cheng. 2008. Responsive supply chain: A competitive strategy in a networked economy. *Omega*, 36(4):549–564.

Gupta, S., S. Modgil and A. Gunasekaran. 2019. Big data in lean six sigma: A review and further research directions. *International Journal of Production Research*, 1–23. doi:10.1080/00207543.2019.1598599.

Handfield, R. and T. Linton. 2017. *The LIVING Supply Chain: The Evolving Imperative of Operating in Real Time*. Hoboken, NJ: Wiley.

Hines, P., M. Holweg and N. Rich. 2004. Learning to evolve: A review of contemporary lean thinking. *International Journal of Operations & Production Management*, 24(10):994–1011.

Jack, E. P. and A. Raturi. 2002. Sources of volume flexibility and their impact on performance. *Journal of Operations Management*, 20(5):519–548.

Jayaram, S., H. I. Connacher and K. W. Lyons. 1997. Virtual assembly using virtual reality techniques. *Computer-Aided Design*, 29(8):575–584.

Kang, H. S., J. Y. Lee, S. Choi, H. Kim, J. H. Park, J. Y. Son and S. D. Noh. 2016. Smart manufacturing: Past research, present findings, and future directions. *International Journal of Precision Engineering and Manufacturing-Green Technology*, 3(1):111–128.

Kaplan, A. and M. Haenlein. 2019. Siri, Siri, in my hand: Who's the fairest in the land? On the interpretations, illustrations, and implications of artificial intelligence. *Business Horizons*, 62(1):15–25.

Kshetri, N. 2017. Blockchain's roles in strengthening cybersecurity and protecting privacy. *Telecommunications Policy*, 41(10):1027–1038.

Kwon, O., N. Lee and B. Shin. 2014. Data quality management, data usage experience and acquisition intention of big data analytics. *International Journal of Information Management*, 34(3):387–394.

Lambert, D. M., M. C. Cooper and J. D. Pagh. 1998. Supply chain management: Implementation issues and research opportunities. *The International Journal of Logistics Management*, 9(2):1–19.

Lamming, R. 1996. Squaring lean supply with supply chain management. *International Journal of Operations & Production Management*, 16(2):183–196.

LaValle, S., E. Lesser, R. Shockley, M. S. Hopkins and N. Kruschwitz. 2011. Big data, analytics and the path from insights to value. *MIT Sloan Management Review*, 52(2): 21–31.

Marodin, G. A., G. L. Tortorella, A. G. Frank and M. G. Filho. 2017. The moderating effect of Lean supply chain management on the impact of Lean shop floor practices on quality and inventory. *Supply Chain Management: An International Journal*, 22(6):473–485.

Martínez-Jurado, P. J. and J. Moyano-Fuentes. 2014. Lean management, supply chain management and sustainability: A literature review. *Journal of Cleaner Production*, 85:134–150.

Monostori, L., B. Kádár, T. Bauernhansl, S. Kondoh, S. Kumara, G. Reinhart, … K. Ueda. 2016. Cyber-physical systems in manufacturing. *CIRP Annals*, 65(2):621–641.

Moyano-Fuentes, J., S. Bruque-Cámara and J. M. Maqueira-Marín. 2019. Development and validation of a lean supply chain management measurement instrument. *Production Planning & Control*, 30(1):20–32.

Nabelsi, V. and S. Gagnon. 2017. Information technology strategy for a patient-oriented, lean, and agile integration of hospital pharmacy and medical equipment supply chains. *International Journal of Production Research*, 55(14):3929–3945.

Ngai, E. W. T., K. K. L. Moon, F. J. Riggins and C. Y. Yi. 2008. RFID research: An academic literature review (1995–2005) and future research directions. *International Journal of Production Economics*, 112(2):510–520.

Olhager, J. and B. M. West. 2002. The house of flexibility: Using the QFD approach to deploy manufacturing flexibility. *International Journal of Operations & Production Management*, 22(1):50–79.

Pfohl, H. C., B. Yahsi and T. Kurnaz. Concept and diffusion-factors of Industry 4.0 in the supply chain. In *Dynamics in Logistics*, edited by M. Freitag, H. Kotzab and J. Pannek, 381–390. Cham: Springer, 2017.

Powell, T. C. and A. Dent-Micallef. 1997. Information technology as competitive advantage: The role of human, business and technology resources. *Strategic Management Journal*, 18(5):375–405.

Qrunfleh, S. and M. Tarafdar. 2013. Lean and agile supply chain strategies and supply chain responsiveness: The role of strategic supplier partnership and postponement. *Supply Chain Management: An International Journal*, 18(6):571–582.

Qrunfleh, S., M. Tarafdar and T. S. Ragu-Nathan. 2012. Examining alignment between supplier management practices and information systems strategy. *Benchmarking: An International Journal*, 19(4/5):604–617.

Reichhart, A. and M. Holweg. 2007. Creating the customer-responsive supply chain: A reconciliation of concepts. *International Journal of Operations & Production Management*, 27(11):1144–1172.

Satoglu, S., A. Ustundag, E. Cevikcan and M. B. Durmusoglu. Lean production systems for Industry 4.0. In *Industry 4.0: Managing The Digital Transformation*, edited by A. Ustundag and E. Cevikcan, 43–59. Cham: Springer Series in Advanced Manufacturing, 2018.

Schmidt, R., M. Möhring, R. C. Härting, C. Reichstein, P. Neumaier and P. Jozinovic. "Industry 4.0 - potentials for creating smart products: Empirical research results". In *Proceedings of the 18th International Conference on Business Information*, edited by W. Abramowicz, 16–27. Poznań: Springer, Cham, 2015. doi:10.1007/978-3-319-19027-3_2.

Scholz-Reiter, B. and M. Freitag. 2007. Autonomous processes in assembly systems. *CIRP Annals of Manufacturing Technology*, 56(2):712–729.

Schrauf, S. and P. Berttram. 2016. Industry 4.0: How Digitization Makes the Supply Chain More Efficient, Agile, and Customer focused. *PwC´s Strategy*, 1–32.

Shah, R. and T. Ward. 2007. Defining and developing measures of lean production. *Journal of Operations Management*, 25(4):785–805.

Stevenson, M. and M. Spring. 2007. Flexibility from a supply chain perspective: Definition and review. *International Journal of Operations & Production Management*, 27(7):685–713.

Swafford, P. M., S. Ghosh and N. Murthy. 2008. Achieving supply chain agility through IT integration and flexibility. *International Journal of Production Economics*, 116(2):288–297.

Swenseth, S. R. and D. L. Olson. 2016. Trade-offs in lean vs. outsourced supply chains. *International Journal of Production Research*, 54(13):4065–4080.

Thomé, A. M. T., L. F. Scavarda and A. J. Scavarda. 2016. Conducting systematic literature review in operations management. *Production Planning & Control*, 27(5):408–420.

Tortorella, G. L. and D. Fettermann. 2018. Implementation of Industry 4.0 and lean production in Brazilian manufacturing companies. *International Journal of Production Research*, 56(8):2975–2987.

Tortorella, G. L., R. Miorando and G. Marodin. 2017. Lean supply chain management: Empirical research on practices, contexts and performance. *International Journal of Production Economics*, 193:98–112.

Tortorella, G., R. Miorando and A. F. Mac Cawley. 2019. The moderating effect of Industry 4.0 on the relationship between lean supply chain management and performance improvement. *Supply Chain Management: An International Journal*, 24(2):301–314.

Tranfield, D., D. Denyer and P. Smart. 2003. Towards a methodology for developing evidence-informed management knowledge by means of systematic review. *British Journal of Management*, 14(3):207–222.

Treiblmaier, H. 2018. The impact of the blockchain on the supply chain: A theory-based research framework and a call for action. *Supply Chain Management: An International Journal*, 23(6):545–559.

Tziantopoulos, K., N. Tsolakis, D. Vlachos and L. Tsironis. 2019. Supply chain reconfiguration opportunities arising from additive manufacturing technologies in the digital era. *Production Planning & Control*, 30(7):510–521.

Vickery, S., R. Calantone and C. Dröge. 1999. Supply chain flexibility: An empirical study. *Journal of Supply Chain Management*, 35(2):16–24.

Vonderembse, M. A., M. Uppal, S. H. Huang and J. P. Dismukes. 2006. Designing supply chains: Towards theory development. *International Journal of Production Economics*, 100(2):223–238.

Want, R. 2006. An introduction to RFID technology. *IEEE Pervasive Computing*, 5(1): 25–33.

Wee, H. M. and S. Wu. 2009. Lean supply chain and its effect on product cost and quality: A case study on Ford Motor Company. *Supply Chain Management: An International Journal*, 14(5):335–341.

Womack, J. P. and D. T. Jones. *Lean Thinking: Banish Waste and Create Wealth in Your Corporation*. New York: Simon and Schuster, 1996.

Womack, J. P., D. T. Jones and D. Roos. *The Machine that Changed the World*. New York: MacMillan/Rawson Associates, 1990.

Index

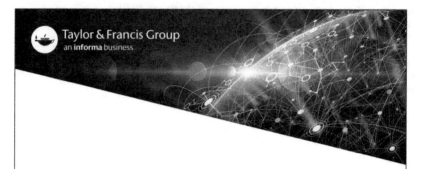